DESIGNING AND CONDUCTING SURVEY RESEARCH

A Comprehensive Guide

THIRD EDITION

Louis M. Rea
Richard A. Parker

JOSSEY-BASS
A Wiley Imprint
www.josseybass.com

Published by Jossey-Bass
A Wiley Imprint
989 Market Street, San Francisco, CA 94103–1741   www.josseybass.com

Jossey-Bass books and products are available through most bookstores. To contact Jossey-Bass directly
call our Customer Care Department within the U.S. at 800–956–7739, outside the U.S. at
317–572–3986, or fax 317–572–4002.

Jossey-Bass also publishes its books in a variety of electronic formats. Some content that appears in print
may not be available in electronic books.

Library of Congress Cataloging-in-Publication Data

Rea, Louis M.
    Designing and conducting survey research: a comprehensive guide/Louis
M. Rea, Richard A. Parker—3rd ed.
        p. cm.
    Includes bibliographical references and index.
    ISBN-13 978-0-7879-7546-3
    ISBN-10 0-7879-7546-X (alk. paper)
    1. Sampling (Statistics)   2. Social surveys.   3. Social
sciences—Statistical methods.   I. Parker, Richard A. (Richard Allen),
    1947–   II. Title.
    HA31.2.R43 2005
    300'.72'3—dc22   2005017233

Printed in the United States of America
THIRD EDITION
HB Printing 10 9 8 7 6 5 4 3 2 1

CONTENTS

FIGURES, TABLES, EXHIBITS, AND WORKSHEETS

Figures

Tables

Exhibits

Worksheets

PREFACE

The sample survey research industry can expect to continue its rapid expansion in the years ahead. As we move through the twenty-first century, myriad technological and analytical innovations have firmly entrenched the probability sample as an indispensable part of life. The growing population and associated socioeconomic complexities, the strengthening of capitalism as a worldwide economic system, and the concurrent forces of democracy surfacing with strength around the globe guarantee the continued significance of sample survey research as a means of gathering data and understanding the interests, concerns, and behavior of people everywhere.

There is a shortage of well-focused, easily understood, yet theoretically and methodologically sound treatments of the sample survey process. Existing texts are generally highly technical and can be appreciated and used only by experts, or they are overly descriptive and not conducive to the successful implementation of a sample survey research project. Furthermore, sample survey research is frequently treated as a relatively small component of broader texts that focus on quantitative methods; this treatment is often insufficiently detailed to serve practitioners in their professional capacities.

At the root of sample survey research is the discipline of statistics. Statistics is an advanced field of study, and traditionally people have had some difficulty mastering it and understanding its wide-ranging applicability to practical problems. One of the most difficult areas of statistics to comprehend is sampling theory. Yet it is specifically through sample surveys that students and researchers most often gather the data they need to carry out their research agendas. Many statistics textbooks do not adequately convey the relationship between sampling theory and its application to the conduct of a survey research project. They tend to inadequately explain the linkage between the theory and its ultimate manifestation in practice.

As teachers and research consultants, we are particularly cognizant of the need to find an appropriate balance between statistical theory and its application. Accordingly, the purpose of this book is to enable readers to conduct a sample survey research project from the initial conception of the research focus to the preparation of the final report, including basic statistical analysis of the data.

Audience

Designing and Conducting Survey Research is intended to serve two distinct audiences. One major audience consists of working professionals who wish to conduct a survey research project or have a framework for planning such a project. Sociologists, political scientists, psychologists, public administrators, city planners, and other social scientists who are faced with gathering data through a survey research project will find this book a useful reference for specific technical and procedural aspects of the survey research project. We also anticipate that professionals in government, private enterprise, and research agencies will make use of this book as a reference guide when the need to conduct, commission, or review sample surveys arises.

An equally important audience is undergraduate and graduate students who want to understand survey research as part of their education in quantitative and research methodology. This third edition of *Designing and Conducting Survey Research* expands on the second edition by incorporating a greater amount of statistical analysis. This permits the book to be a more complete guide for designing, conducting, and analyzing sample survey research. The book not only serves as a comprehensive guide to the survey research process but also provides many more useful statistical techniques than the second edition did. The exercises and examples throughout this edition have been expanded to span a multitude of subjects, thereby appealing to a broad-based audience in the social and behavioral sciences.

Overview of the Contents

Part One covers a major component of the survey research process: construction of the survey instrument. Chapter One presents an overview of the sample survey research process, including the advantages and disadvantages of the major types of survey research: mail-out, telephone, Web based, intercept, and in-person interview surveys. The procedures for the administration of these surveys are also discussed.

Chapters Two and Three pursue in detail the process of questionnaire development. Drawing on specific survey research projects we have conducted, we demonstrate the key components and principles associated with questionnaire design. Chapter Two provides an overview of the process, including thorough discussions of question type and sequence. Chapter Three delves into the specific guidelines for the construction of individual questions and deals with such topics as phraseology, format, and, in particular, the avoidance of bias-inducing questions.

Chapter Four discusses the role of focus groups in preparing survey questionnaires and analyzing survey findings. The focus group process is presented in four stages: planning, recruiting, conducting, and analyzing. This chapter recognizes the rapidly growing importance of focus group research and draws on our own extensive experience in this area of research.

Part Two addresses the more technical aspects of the survey research process, from its theoretical underpinnings to scientific sample selection procedures. Chapter Five establishes the groundwork for the ensuing chapters through a treatment of descriptive statistics, including measures of central tendency, measures of dispersion, and the normal distribution. Chapter Six explains how generalizations about an entire population can be made from just one sample consisting of a relatively small subset of that total population. Chapter Seven presents practical applications of the theory explained in the preceding chapter in terms of confidence intervals and the testing of simple hypotheses based on a sample. Chapter Eight thoroughly discusses the important considerations associated with determining an appropriate sample size for the conduct of sample surveys. It presents and explains equations for determining sample sizes and also includes tables that demonstrate how the required sample size can be referenced quickly.

Chapters Six, Seven, and Eight are particularly valuable contributions to the survey research literature. Their relevance is derived from a clear and focused presentation of complex material that historically has represented a difficult obstacle for students and practitioners to overcome.

Chapter Nine introduces the concepts of general population, working population, and sampling frame. Procedures for selecting probability samples are pursued in detail, and various methods of probability and nonprobability sampling are explored.

Part Three explores the analysis of survey findings, including basic statistical techniques. It presents the statistical techniques most commonly used in testing the statistical significance of survey data. It addresses the question, "Are the apparent relationships in the survey data, as displayed in tables, genuine, or are they the result of chance occurrences?" Chapter Ten, referencing our personal research experience, presents a thorough discussion of the appropriate use of cross-tabulated tables and introduces the chi-square test of significance along with Cramér's V, gamma, and lambda.

Chapter Eleven discusses independent and paired samples t tests, difference of proportions, and analysis of variance, and Chapter Twelve addresses regression and correlation. A great deal of the information presented in Chapters Eleven and Twelve is new in this edition, and this information contributes substantially to the analysis and presentation phases of the survey research process.

Chapter Thirteen provides rules and guidelines for the preparation of final research reports. It focuses on the graphical depictions of data and the appropriate use and organization of tables, while discussing the merging of text and statistical information into an integrated work that effectively communicates a study's findings to its intended audience.

Years of teaching statistics and survey research as well as extensive experience in private consulting motivated us to prepare a book that could serve both practitioners and academics. It is this combination of classroom and field experience that gives this work its unique approach and instructional value. Seven years of working with the second edition prompted us to supplement, clarify, and refine certain aspects of the book in order to serve our audience better. We trust that readers will find this edition to be useful and highly conducive to their research endeavors.

San Diego, California   Louis M. Rea
July 2005   Richard A. Parker

THE AUTHORS

Louis M. Rea is professor of city planning and director of the School of Public Administration and Urban Studies at San Diego State University. He received his B.A. degree in economics from Colgate University and both his M.R.P. degree and Ph.D. degree in social science from Syracuse University.

Rea has taught graduate courses in statistical analysis, transportation planning, survey research, and urban and fiscal problems. He has had extensive experience as a researcher and consultant in the San Diego area since 1975 and has conducted surveys in numerous consulting and research assignments for municipal jurisdictions and private businesses throughout southern California. He also has prepared environmental impact reports and market analyses for various commercial and recreational developments. Among other projects, he has analyzed the feasibility of assessment districts and direct benefit financing, conducted research in the area of transportation, and prepared demographic and economic profiles and projections for numerous public and private agencies.

Rea has published a variety of articles in such journals as *Urban Affairs Quarterly, Transportation Quarterly,* and the *American Review of Public Administration.* He has participated in panel discussions and has delivered numerous papers at professional conferences throughout the United States.

Richard A. Parker is professor in the School of Public Administration and Urban Studies at San Diego State University, where he teaches courses in urban economic development, survey research, and statistical methodology. He received his B.S. degree in business administration and his M.B.A. degree from the University of California, Berkeley; his Master of City Planning degree from San Diego State University; and his Ph.D. degree in business administration from Pacific Western University.

Parker is a survey and market research and economics consultant to both the public and private sectors. He specializes in survey research for housing, retail, commercial, recreational, and transportation development and for environmental, socioeconomic, demographic, and fiscal impact analyses. He has been involved in a number of projects concerning redevelopment and growth in southern California and has published articles in the *Glendale Law Review* and the *Western Governmental Researcher* and a monograph published by the University of California Center for Real Estate and Urban Economics. He has also delivered papers at conferences in the field of urban development and fiscal impact. Furthermore, he has presented survey research and focus group studies at conferences in the southwestern United States.

Parker possesses extensive analytical experience in real estate and real estate investment, having served for many years as director of real estate operations and investments for a major southern California business management firm before returning to academia in 1982.

Together, Parker and Rea prepared the first fiscal impact analysis of the provision of public services to undocumented immigrants in the state of California. The analysis included extensive survey research, focus group discussions, and the use of advanced statistical techniques.

The authors have also conducted extensive on-board bus and rail surveys for some of the largest transportation authorities in the United States. These projects have required the use of advanced sampling techniques and the application of survey research principles that inform and benefit this edition of the book.

DESIGNING AND CONDUCTING SURVEY RESEARCH

PART ONE

DEVELOPING AND ADMINISTERING QUESTIONNAIRES

CHAPTER ONE

AN OVERVIEW OF THE SAMPLE SURVEY PROCESS

Surveys have become a widely used and acknowledged research tool in most of the developed countries of the world. Through reports presented by newspapers, magazines, television, and radio, the concept of considering information derived from a relatively small number of people to be an accurate representation of a significantly larger number of people has become a familiar one. Surveys have broad appeal, particularly in democratic cultures, because they are perceived as a reflection of the attitudes, preferences, and opinions of the very people from whom the society's policymakers derive their mandate. Politicians rely heavily on surveys and public opinion polls for popular guidance in mapping out campaign strategies and carrying out their professional responsibilities. Commercial enterprises use survey findings to formulate market strategies for the potential widespread use, distribution, and performance of new and existing products. Television and radio programs are evaluated and scheduled largely in accordance with the results of consumer surveys. Government programs designed to provide assistance to various communities often rely on the results of surveys to determine program effectiveness. Private social organizations obtain information from their members through the use of survey techniques. Libraries, restaurants, financial institutions, recreational facilities, and churches, among many others, make use of polls to solicit information from their constituents and clientele concerning desired services.

As a research technique in the social sciences and professional disciplines, survey research has derived considerable credibility from its widespread acceptance and use in academic institutions. Many universities have established survey research institutes where the techniques of survey research are taught and surveys can be conducted within the confines of propriety and scientific rigor. Students are often encouraged to use survey research for gathering primary data, thereby satisfying the requirement of conducting original research. Professors publish countless articles and books based on the results of funded and unfunded survey research projects.

Despite the broad-based societal acceptance of survey research, there remains a lingering doubt, especially within the general population, at large, concerning the reliability of information derived from relatively few respondents purporting to represent the whole. They frequently ask, for instance, "How can fifteen hundred respondents to a survey be said to represent millions of people?" or "Why should two thousand television viewers dictate to program directors on a national scale what Americans choose to watch?" The answers to these and other such questions lie in the systematic application of the technique of scientific sample survey research.

Survey research involves soliciting self-reported verbal information from people about themselves. The ultimate goal of sample survey research is to allow researchers to generalize about a large population by studying only a small portion of that population. Accurate generalization derives from applying the set of orderly procedures that comprise scientific sample survey research. These procedures specify what information is to be obtained, how it will be collected, from whom it will be solicited, and how it will be analyzed.

If the researcher needs personal, self-reported information that is not available elsewhere and if generalization of findings to a larger population is desired, sample survey research is the most appropriate method as long as enough general information is known or can conveniently be obtained about the subject matter under investigation to formulate specific questions and as long as the population that is needed to be sampled is accessible and willing to provide self-reported information. The theoretical underpinnings of scientific sample survey research, its procedural applications, and analysis of the data it generates constitute the substance of this book.

Gathering Information Through Research

The researcher must be aware that survey research is only one among several methods associated with the process of data collection. The three main techniques used to collect primary data (data collected firsthand, directly from the subjects under study) are survey research, direct measurement, and observation, all of which in one way or another can make use of sampling. Secondary research is a fourth means of data collection. It consists of compiling and analyzing data that have already been collected and exist in usable form. These alternative techniques, when they are not appropriate in and of themselves, can often be used as complements to the survey research process. A brief description of these alternative techniques follows:

- *Secondary research:* Certain data may already exist that can serve to satisfy the research requirements of a particular study. Any study should investigate existing sources of information as a first step in the research process to take advantage of information that has already been collected and may shed light on the study. Sources of secondary information include libraries, government agencies, and private foundations, among others.
- *Direct measurement:* This technique involves testing subjects or otherwise directly counting or measuring data. Testing cholesterol levels, monitoring airport noise levels, measuring the height of a building to make certain it complies with local ordinances, and counting ballots in a local election are all examples of direct measurement.
- *Observation:* A primary characteristic of observation is that it involves the direct study of behavior by simply watching the subjects of the study without intruding on them and recording certain critical natural responses to their environment. For example, a government official can obtain important information about the issues discussed in a speech by observing the audience's reactions to that speech.

However, there is no better method of research than the sample survey process for determining, with a known level of accuracy, detailed and personal information about large populations. Opinions, which are the keys to public policy, are obtainable with defined and determinable reliability only through the survey research process. By combining surveys with scientific sampling, the researcher is using the only method of gaining this information to a known level of accuracy. The survey process is particularly suited to collecting data that can inform the researcher about research questions such as the following:

- How do Americans feel about proposed changes in social security regulations?
- What is the average income of people twenty-five years of age and older whose highest level of completed education is high school?
- What factors influence people's choice of banks?

- What are the reactions among employees of a local factory concerning a newly proposed union policy?
- How do members of the New York State Bar Association feel about capital punishment?
- What do various state legislators think about a proposed mandatory balanced-budget amendment?
- What proportion of drivers observe seat belt laws?
- To what extent has the Latino community in Texas experienced job discrimination?

The particular use for which a survey is conducted determines the informational requirements of that survey. Surveys typically collect three types of information: descriptive, behavioral, and attitudinal.

Surveys frequently include questions designed to elicit descriptive information or facts about the respondent. Such important data as the respondent's income, age, education, ethnicity, household size, and family composition are integral to most sample survey studies. These socioeconomic characteristics provide important information that enables the researcher to better understand the larger population represented by the sample.

In many survey research projects, the researcher is interested in the respondent's behavior. Patterns of transportation use, recreation, entertainment, and personal behavior are often the desired information in sample survey studies. For example, such information as frequency of public transit ridership or use of various types of recreational and entertainment facilities is typical of behaviorally oriented information that can be obtained from sample surveys.

In addition to descriptive and behavioral data, many surveys solicit, as their primary focus, the respondent's attitudes and opinions about a variety of conditions and circumstances. The hallmark of this type of sample survey is the public opinion poll, which seeks opinions and preferences regarding issues of social and political relevance. The primary objective of such studies is to be predictive and future oriented.

Very rarely does a study include only one of the above informational categories. Scientific investigation requires that relationships be identified in terms of descriptive, behavioral, and attitudinal data so that we may fully understand the differential complexities of the population from which a sample has been drawn. For instance, in a political public opinion poll, it is much more desirable to know not only the breakdown of votes for each candidate but also such factors as the voter's political party, age, gender, past voting patterns, and opinion on a variety of key issues. Such a survey requires the researcher to derive information from each of the above categories in one sample survey.

Advantages of Sample Survey Research

Generalizations based on a mere fraction of the total population (a sample) did not gain acceptance until the beginning of the twentieth century, when a researcher for a liquor distillery in England named W. S. Gossett was faced with the problem of testing the quality of his company's product. Testing the plant's output involved tasting, and therefore consuming, the product. Testing the entire output of the plant, or even as few as one in ten bottles, was clearly not economically feasible. Gossett, writing under the pseudonym "Student," therefore developed a theoretical basis for making generalizations about the quality of the plant's product by sampling only a small portion of that output.

The foremost advantage of the sample survey technique, as indicated by Gossett's experience, is the ability to generalize about an entire population by drawing inferences based on data drawn from a small portion of that population. The cost and time requirements of conducting a sample survey are significantly less than those involved with canvassing the entire population. When implemented properly, the sample survey is a reasonably accurate method of collecting data. It offers an opportunity to reveal the characteristics of institutions and communities by studying individuals and other components of those communities that represent these entities in a relatively unbiased and scientifically rigorous manner.

Surveys can be implemented in a timely fashion. That is, the survey project can be organized so that the actual data gathering is performed in a relatively short period of time. Besides the convenience afforded by this approach, there is also the advantage of obtaining a snapshot of the population. Other techniques may involve a longer-term study, during which opinions or facts may change from the beginning of the study to the end.

Well-structured sample surveys generate standardized data that are extremely amenable to quantification and consequent computerization and statistical analysis. This quality has been enhanced through rapid advances in computer technology as well as through the development and refinement of complex analytical statistical software packages and techniques. For purposes of comparisons among individuals, institutions, or communities, surveys offer a further advantage: replicability. A questionnaire that has been used in one city or community can be reimplemented in another community or administered once again in the same community at a later date in order to assess differences attributable to location or time.

Sampling started gaining general acceptance beginning in 1935, when George Gallup established the American Institute of Public Opinion in order to conduct weekly polls on national political and consumer issues for private and public sector clients. Inasmuch as Gallup was operating a business for profit and since he was committed to delivering weekly polls, he was necessarily highly sensitive to cost and time factors. Gallup developed a method of sampling fifteen hundred to three thousand respondents—quite a small number compared to other surveys at that time. His method involved establishing sample quotas based on age, sex, and geographical region. In the 1936 presidential election between Franklin D. Roosevelt and Alfred Landon, Gallup forecast a Roosevelt victory, while one of the most respected polls at that time, the Literary Digest poll of 2.5 million subscribers, forecast a Landon landslide. The final results are well known: a Roosevelt victory with 61 percent of the vote. The scientifically implemented small sample thereafter became established as the survey method of choice. Advancements in the understanding of sample survey methodology that were developed in World War II and refined thereafter now provide even greater accuracy than Gallup had in 1936, with still smaller sample sizes.

Types of Sample Survey Research

Survey information can be collected by means of any of five general methods of implementation: mail-out, Web based, telephone, in-person interviews, and intercept. This section addresses the advantages and disadvantages of these types of surveys and discusses the procedures for administering the surveys.

Mail-Out Surveys

The mail-out format for collecting survey data involves the dissemination of printed questionnaires through the mail (commonly the postal service) to a sample of predesignated potential respondents. Respondents are asked to complete the questionnaire on their own and return it by mail to the researcher.

Advantages and Disadvantages. The advantages of the mail-out technique can be stated as follows:

- *Possible cost savings:* Other techniques require trained interviewers, and the recruitment, training, and employment of interviewers can be quite costly. Access to respondents by mail can, under some circumstances, be less expensive than telephone surveys and certainly less expensive than in-person interviews.
- *Convenience:* The questionnaire can be completed at the respondent's convenience.

- *Ample time:* The respondent has virtually no time constraints. There is enough time to elaborate on answers and consult personal records if necessary to complete certain questions.
- *Authoritative impressions:* The researcher can prepare the mail-out questionnaire form so that it has significant legitimacy and credibility.
- *Anonymity:* Because there is no personal contact with an interviewer, the respondent may feel that the responses given are more anonymous than is the case with other formats.
- *Reduced interviewer-induced bias:* The mail-out questionnaire exposes all respondents to precisely the same wording on questions. Thus, it is not subject to interviewer-induced bias in terms of voice inflection, misreading of the questions, or other clerical or administrative errors.
- *Complexity:* Mail-out questions can be longer and more complex than telephone questions.
- *Visual aids:* Mail-out questionnaires can make use of photographs and maps that would be impossible to utilize in a telephone survey.

Mail-out questionnaires have certain disadvantages, however, which can be summarized as follows:

- *Comparatively long time period:* Many follow-ups and substitutions of sample respondents are required in order to achieve the appropriate sample size and adequate random distribution necessary for purposes of generalization. The mail-out, therefore, generally requires a few weeks for questionnaires to be returned.
- *Self-selection:* Mail-outs typically achieve a lower response rate than telephone surveys. Low response rates can imply some bias in the sample. For instance, poorly educated respondents or those with reading or language deficiencies tend to exclude themselves from this form of survey more often than from surveys administered by an interviewer.
- *Lack of interviewer involvement:* The fact that no interviewer is present means that unclear questions cannot be explained, there is no certainty that the questions will be answered in the order written (which may be important), and spontaneously volunteered reactions and information are not likely to be recorded by the respondent and cannot be probed by an interviewer as would be the case with other methods.
- *Incomplete open-ended questions:* It is more likely that questions requiring an original written response in lieu of fixed answers will be avoided.

Administration of Mail-Out Surveys. Certain guidelines should be followed in administering a mail-out questionnaire. First, the questionnaire should be designed in the form of a booklet in order to ensure a professional appearance and to make it more usable by the respondent. Any resemblance to an advertising brochure should be strictly avoided. The aesthetic appearance of the questionnaire is important in terms of generating satisfactory response rates. There should be adequate spacing between questions, and questions should not begin on one page and end on another. Instructions to the respondent should be clear and easily distinguished from the survey questions themselves. Graphics, such as maps and illustrative photographs, should be carefully integrated into the design of the questionnaire.

The cover letter should be prepared in accordance with the principles discussed later in this chapter and should become the first page of the booklet. The last page of the booklet should be reserved for three purposes only: to express appreciation to the respondents for their participation, to provide a return mailing address and prepaid postage through a business reply permit, and to provide instructions for returning the completed questionnaire. Alternatively, it is possible to provide postage-paid, preaddressed return envelopes, but the cost of this approach is somewhat higher.

Questionnaires should be stamped with an identification number or barcode for purposes of monitoring the follow-up process. This number or code must be explained to the respondent in the cover letter, accompanied by assurances of privacy and confidentiality.

The questionnaire booklet is mailed by first-class postage to the respondent in an envelope. The envelope is addressed either with the name and address of the respondent individually imprinted (in the case of small, more personalized surveys) or with a mailing label (most commonly used in large-scale, high-volume surveys). A target date should be designated for the return of the questionnaire; this target date is generally recommended to be approximately two weeks from the initial mailing date. Two weeks after the initial mailing, a follow-up postcard reminder should be sent to potential respondents who have not yet replied, as determined by their prestamped identification number. The reminder should be friendly in tone and indicate that if the completed questionnaire and the reminder postcard have crossed in the mail, the respondent should disregard the reminder; it should also again express appreciation for the respondent's cooperation.

Four weeks from the initial mailing, a second follow-up can be mailed to all survey recipients who have not yet responded. This follow-up should include a new cover letter that does not specify a target due date but instead stresses the importance of responding. Another copy of the questionnaire should accompany the letter in case the original questionnaire has been misplaced or discarded.

It can be reasonably expected that this procedure will yield a response rate that can approach 50 percent for the general public and a somewhat higher rate for

specialized populations. The researcher should wait two weeks after the second follow-up before closing the mailing process. A response rate of 50 percent can be considered satisfactory for purposes of analysis and reporting of findings as long as the researcher is satisfied in the representativeness of the respondents (see Chapter Nine). If the researcher wishes to increase the response rate and has adequate resources and time to do so, the following additional procedures are suggested:

- In lieu of using mailing labels, envelopes and the cover letter can be individually imprinted with the potential respondent's name and address.
- The cover letters should be individually signed in blue ink to avoid the impression that they were impersonally mass-produced.
- The follow-up mailings should include eye-catching but tasteful illustrations and graphics.
- Six weeks after the initial mailing, nonrespondents can be given a reminder telephone call.
- A third follow-up mailing, again with a new cover letter and copy of the questionnaire, can be sent to all nonrespondents eight weeks after the first mailing. This third follow-up should be delivered by certified mail.

These additional procedures are designed to achieve a response rate in excess of 70 percent for the general population and as high as 90 percent for certain specialized groups.

Web-Based Surveys

The Web-based survey is an alternative to the traditional mail-out technique whereby individuals are contacted by e-mail and asked to participate in a survey that is designed to be completed and submitted by computer.

Advantages and Disadvantages. The advantages of the Web-based survey method are as follows:

- *Convenience:* This technique represents a convenient and efficient way of reaching potential respondents. They are able to receive the questionnaire and complete it in the privacy of their home or office. This advantage is becoming particularly significant as the availability of computers becomes increasingly widespread.
- *Rapid data collection:* Information, especially information that must be timely (for instance, a political public opinion poll related to an upcoming election), can be collected and processed within days.

- *Cost-effectiveness:* This technique is more cost-effective than the traditional mail-out survey because there is no need for postage or paper supplies. It is also more cost-effective than the telephone and in-person surveys because it is not at all labor intensive.
- *Ample time:* The respondent is not pressed for time in responding to the Web-based survey and has the opportunity to consult records in answering the questions. There is time to consider response choices and to respond to open-ended questions in the form of text.
- *Ease of follow-up:* Potential respondents can be reminded to respond to the Web-based survey through follow-up e-mail messages.
- *Confidentiality and security:* Personal or sensitive information supplied by the respondents can be protected on a secure server through the efforts of the research team.
- *Specialized populations:* The Web-based survey is particularly useful in reaching specialized or well-identified populations whose e-mail addresses are readily available. For example, we have successfully used this technique to conduct surveys of satisfaction among employees and stakeholders of large public organizations.
- *Complexity and visual aids:* As with mail-out surveys, Web-based surveys can utilize visual images and more complex questions.

Web-based surveys have certain disadvantages that can be delineated as follows:

- *Limited respondent bases:* A major disadvantage of this technique is that it is limited to populations that have access to e-mail and a computer. Furthermore, the technique assumes a certain minimal level of computer literacy that is necessary for the completion and submission of the questionnaire. Such literacy is improving within the general population but is hardly widespread.
- *Self-selection:* As in the traditional mail-out, there is a self-selection bias that leads to lower response rates. Those who do not use e-mail or are not comfortable with Web-based technology exclude themselves from the sample. Also, individuals with reading or language issues tend not to respond to Web-based surveys. Some researchers send the survey by e-mail in multiple languages in an effort to obviate this problem.
- *Lack of interviewer involvement:* Since there is no interviewer involvement in the Web-based survey, unclear questions cannot be explained, and respondents may not follow instructions. These problems can seriously compromise the scientific reliability of the survey even though telephone contacts are provided to the respondents in the event that they need help.

Administration of Web-Based Surveys.

The Web-based survey is not complex to administer, especially in comparison to the traditional mail-out technique. It is critical that at least one member of the research team be thoroughly conversant with Web-based technology. Assuming the research team has the necessary technical expertise and knowledge, the following administrative tasks are required to implement a Web-based survey successfully:

- The researchers must have access to a listserv that comprises the e-mail addresses of the sample respondents. The researchers should be able to send e-mail messages to everyone simultaneously as well as to selected individuals as desired and necessary.
- The initial e-mail message to all sample respondents is designed to invite participation in the Web-based survey by providing the potential respondent with a unique password that protects against multiple responses by the same respondent. This password enables the respondent to gain access to the questionnaire by clicking on a URL provided with the e-mail.
- The researcher must provide clear instructions so the respondent understands how to navigate through the questionnaire and submit answers to the researcher. The researcher should strive to prepare a questionnaire that is as user friendly as possible.
- It is important that the questionnaire be submitted to a secure server so that the privacy of the respondents is maintained.
- The researcher should be able to apply a statistical program (such as SPSS) to the data and prepare the necessary tables, charts, and statistical analyses.

The Telephone Survey

The telephone survey is a method of collecting information through the use of telephone interviews between a trained interviewer and selected respondents.

Advantages and Disadvantages.

The advantages of the telephone survey interviewing process can be stated as follows:

- *Rapid data collection:* Information, especially information that must be timely (for instance, a political public opinion poll related to an upcoming election), can be collected and processed within days. It is possible to complete a telephone survey in the time it would take simply to plan a mail-out or in-person survey.
- *Possible cost savings:* The cost of implementing a telephone survey is considerably less than that of in-person interviews, and under certain circumstances, it can be less than that of a mail-out survey.

- *Anonymity:* A telephone survey is more anonymous than an in-person interview. Hence, the interviewer can conduct in-depth questioning in a less-threatening environment than exists in face-to-face situations.
- *Assurance that instructions are followed:* As with the in-person interview, the telephone interviewer can make certain that the questions are answered in precisely the order intended so that the integrity of the questionnaire sequence is maintained.

Telephone surveys also have certain disadvantages:

- *Less control:* The interviewer has less control over the interview situation in a telephone survey than in an in-person interview. The respondent can easily end the interview at any time by hanging up the telephone.
- *Less credibility:* The interviewer will have greater difficulty establishing credibility and trust with a respondent over the telephone than would be the case in person or by mail.
- *Lack of visual materials:* Both the mail-out survey and the in-person interview permit the use of visual aids, such as maps, pictures, or charts, as components of the questions. The telephone survey does not provide such an opportunity to the researcher.
- *Less complexity:* Related to the lack of visual materials is the fact that telephone questions must be much less complex than most other survey forms.

Administration of Telephone Survey. The telephone survey is less complex to implement than the mail-out. The most important aspect of this survey technique is the use of personal interviewers; the proper selection and training of these interviewers is critical to the success of the research project.

Selection of Telephone Interviewers. The researcher should be aware of the fact that there are a variety of sources through which individuals may be recruited to serve as telephone interviewers. The single best source of interviewers, when available, is a local university. Students, especially upper-division undergraduate students and graduate students, are motivated to become involved in the interviewing process for two basic reasons. First, there is frequently some substantive interest in the research project and its potential findings. Second, students often seek ways to augment their income to help fund their education while at the same time gaining relevant experience and therefore may be willing to work for wages that are relatively modest in relation to their skill level. If the researcher does not already have an affiliation with a university, professors in appropriate disciplines should be contacted and arrangements made to recruit potential interviewers. University bulletin boards and newsletters can also be used. If universities are not easily available, temporary agencies can be a source of employees, and when universities and temporary agencies are not readily accessible or when additional assistance is required, newspaper "help wanted" ads are the next most effective recruitment tool. Newspapers that can be considered for placement of such ads include not only the major metropolitan dailies but also neighborhood weekly newspapers. Another source of recruitment is contact with local organizations such as social service delivery groups, civic organizations, and church groups, which are frequently able to publicize recruitment needs among their memberships.

The content of the recruiting advertisement should enable potential applicants to determine if they are interested in the job and if they meet its requirements. Thus, the job notice should include such information as work hours, pay rate, location of the work site (home or central telephone facility), and whether fluency in a language other than English is necessary. The job notice should also indicate times and dates for group meetings, which are designed to dispense additional information, answer questions, and receive interviewer job applications; these applications should contain questions about work history, education, professional references, and availability to perform the required tasks. Group sessions are an efficient way to avoid unscheduled and frequent individual recruitment sessions, which can be time-consuming for the researcher.

Having reviewed the job applications, the researcher should narrow the list of applicants by screening out those who clearly do not meet the basic requirements. After a brief personal interview, the remaining applicants are asked to administer a practice questionnaire as a final screening device. This process will enable the researcher to determine the applicants' ability to read at the appropriate level, follow directions, and relate to other people. Final selection should be based on the written application, the personal interviews, the practice questionnaire, and any potential biasing characteristics that the interviewer feels the applicant may possess. A poor performance during the practice questionnaire should not necessarily eliminate the applicant from consideration; interviewer training after selection may help to mitigate some of the problems that are seen during the practice session.

Training of Telephone Interviewers. Interviewer training consists of a two-pronged process. The researcher first provides the interviewer with general training regarding the fundamental techniques of the interviewing process and then instructs the interviewer in proper administration of the specific survey questionnaire. Several procedures can be used to assist in the training process. To begin, an overview of the questionnaire should be provided that is specific to the study, with the various types of questions identified and all interviewer instructions pointed out, especially those pertaining to filtering and screening. It is advisable to pay particular attention to questions that permit more than one response and to make certain that "Other" categories and open-ended questions are recorded with precision. The researcher should also discuss the answer code format and explain the purpose of the variable fields.

Interviewers should be provided with a general understanding of the scope and substantive purpose of the research project. The organization sponsoring the survey should also be indicated. It is also important to make interviewers aware of the role they play within the survey process as a whole; that is, the interviewers should become aware of the sample size, the sample selection process employed, and how their role relates to the entire survey process, including data entry, data analysis, and the preparation of the final report.

The interviewer should be given the opportunity to practice administering the questionnaire. Before actual interviews begin, interviewers should meet with the researcher for a final rehearsal of the questionnaire. All interviewers should be present, and they should alternate the roles of interviewer and respondent.

All telephone interviewers should be aware of some general ethical issues. The interview must be held in confidence, and any information obtained through the interviewing process must be treated anonymously. The interviewer achieves the proper degree of confidentiality and anonymity by making no notations on the survey forms that would permit identification of the respondent. The telephone survey process permits the researcher to note immediately who has responded and who has not. The researcher will provide to the interviewer a sample list of telephone numbers, on which the interviewer should make the appropriate notations. In particular, when an interview has been completed, the corresponding telephone number should be deleted from the list. In contrast, in mail-out follow-up surveys, the researcher must identify each returned questionnaire. Although ethics demand anonymity and confidentiality in both formats, the proper application of the telephone interview provides a built-in safeguard that the mail-out survey does not possess.

The interviewer must be careful to minimize the amount of bias introduced into the interviewing process. The introductory greeting, as discussed in Chapter Two, should be delivered with sincerity. Questions should be read verbatim with appropriate pacing and in a pleasant conversational tone. The interviewer should be satisfied that the respondent understands the question and must be careful to record responses accurately, making certain that the respondent's answer is fully understood.

The interviewer should not express any opinions or make extraneous comments in reaction to statements made by the respondent. Despite these efforts to minimize bias, there is always the potential for the respondent's answers to be affected to some extent by her or his reaction to one or more characteristics of the interviewer, such as ethnic or regional accents, sex, or age. The researcher should be cognizant of these potential problems and plan the conduct of the research study accordingly.

Interviewers should input all responses directly onto the computerized questionnaire form. Direct use of the form in its electronic format makes it considerably easier for the interviewer to follow all the instructions and ask all the relevant questions, especially when filtering or screening questions are used. If the respondent offers extraneous or supplementary information, the interviewer should be instructed to record it as accurately as possible. Such voluntary statements may contain valuable information that may shed light on the issue at hand.

Interviewing should be conducted in the early evening (6:00 P.M. to 9:00 P.M. local time) and on the weekends (noon to 9:00 P.M.). Evenings provide the interviewer greater opportunity to reach working adult household members, whereas daytime calling during the week reaches only adults who are not working outside the home. After 9:00 P.M., the interviewer should stop placing calls to avoid disturbing those who may have retired for the night. Similarly, on weekends, calls prior to noon may interfere with needed extra hours of sleep or time spent at religious services. The overriding principle is to reach as many adult household members as possible at a convenient time.

If the interviewer encounters a busy signal, the call should be tried again in thirty minutes; if the line is still busy, the call should be placed again the next day. If the first call on the next day is once again met with a busy signal, the interviewer should again wait thirty minutes and try one more time. When there are repeated busy signals, the interviewer is required to contact the telephone company to ascertain the working status of the number. If the telephone company indicates that the line is operating, the interviewer may try calling on another day at a time totally different from the previous attempts. If the line is still busy, the interviewer should classify the number as "nonresponse" to avoid spending an inordinate amount of time in pursuit of one potential respondent. When, instead of a busy signal, the first call elicits no answer, the call should be repeated the next day. If there is still no answer after four or five such attempts, the telephone number can be treated as a "nonresponse."

When the sample is comprised of households, rather than specific individuals at a given phone number, the interviewer must speak to an adult member of the selected household unless the survey is specifically geared to minors. The interviewer should try to speak to a representative mix of men and women and sometimes may have to specifically request to speak to an "adult male" or "adult female" in order to maintain representativeness by gender.

When the interviewer has exhausted the sample list of telephone numbers, he or she should tell the researcher how many nonresponses have been encountered. The researcher will provide the interviewer with a list of replacement telephone numbers selected in accordance with the appropriate sample selection method (see Chapter Nine). The interviewer then proceeds to make these calls as described above, returning to the researcher once again all nonresponses from the list. This process continues until the interviewer has completed the number of interviews assigned.

At the completion of each interview, interviewers should examine the completed questionnaire for missed questions, unclear open-ended responses, and general readability. If necessary, a follow-up telephone call to the respondent should be conducted immediately.

There are a number of additional rules of interviewing that the researcher should insist on having followed. These rules include the following:

- An interviewer should never interview more than one adult in the same household.
- A friend or relative of the research team should not be interviewed. If a friend or relative is part of the sample list, the researcher should be notified so that the person in question can be reassigned to another interviewer.
- The interviews should be conducted in as much privacy as possible to avoid distraction.
- The interviewer should not delegate assigned interviews to anyone else unless for reasons of language.
- Interviews should never be falsified.

In-Person Interviews

In-person, or face-to-face, surveys are structured to permit an interviewer to solicit information directly from a respondent in personal interviews.

Advantages and Disadvantages. The advantages of the in-person interview survey technique are as follows:

- *Flexibility:* The interviewer can probe for more detail, explain unclear questions, and use visual aids, such as maps or photographs.
- *Greater complexity:* Interviewers can administer highly complex questionnaires and provide detailed instructions and lengthy lists of alternative responses that many respondents would find confusing and intimidating if the questionnaire were administered by any other means.

- *Ability to contact hard-to-reach populations:* Certain groups, for instance, the homeless or criminal offenders, are difficult or impossible to reach by any method other than personal interviews.
- *Assurance that instructions are followed:* The interviewer can make certain that the questions are answered in precisely the order intended so that the integrity of the questionnaire sequence is maintained.

In-person interviews also have certain disadvantages:

- *High cost:* Administering in-person interviews can be very costly in terms of time per interview, travel time, interviewer training, and field supervision.
- *Interviewer-induced bias:* Although the interviewer obviously serves many useful functions in this process, he or she can also be a source of bias. For example, the interviewer may inadvertently react in some way to a response rather than remaining neutral. This action could affect future responses by the interviewee and, hence, the validity of the entire questionnaire. By the same token, the respondent may alter his or her responses to gain perceived approval from the interviewer.
- *Respondents' reluctance to cooperate:* If respondents must allow interviewers into their homes to participate in a face-to-face survey, they may tend to be somewhat less inclined to participate than in a telephone survey. Many telephone calls and return visits may be necessary in order to complete an interview.
- *Greater stress:* The in-person interview format is clearly the most intense and stressful for both the respondent and the interviewer. It tends to be a longer and more complex interviewing process, and it is the only one in which a stranger is present in the respondent's environment. Such situations can cause increased stress and fatigue, which may have unfavorable effects on the quality of the responses.
- *Less anonymity:* The advantages of the anonymity perceived by the respondent in mail-out and telephone surveys are greatly reduced in the face-to-face format.

Administration of In-Person Interviews. The administration of in-person, or face-to-face, interviews creates even greater challenges for the researcher than do other methods. As with the telephone interview, the selection and training of interviewers is critical to the successful solicitation of data.

Selection of In-Person Interviewers. The process of selecting in-person interviewers should be precisely the same as that used for selecting telephone interviewers, with a certain emphasis on physical characteristics that is not as important in the telephone survey process. Because in-person interviewing involves face-to-face interaction between the respondent and the interviewer, the respondent's willingness to participate is highly dependent on the comfort level the respondent perceives. Physical characteristics such as attire, cleanliness, neatness, manners, and overall grooming loom considerably larger in the in-person format than in the telephone survey, and they set the tone for the seriousness of the research study. Consequently, these characteristics must be emphasized in the selection process.

There is a secondary component of the interviewer's physical characteristics that can bear strongly on the in-person interview. A series of studies throughout the years has established that people have been socialized to react differently to another person depending on his or her sex, age, ethnicity, and social status. These studies indicate that an interviewer with roughly the same characteristics as the respondent will tend to obtain more reliable information, especially if this information pertains to issues that the respondent perceives are sensitive in nature. In the interest of obtaining as much reliable information as possible, the researcher must incorporate these considerations into the interviewer selection process.

Training of In-Person Interviewers. The principles of interviewer training that have been stated with regard to the training of telephone interviewers apply also to in-person interviewers. A few additional considerations exist, a result of the differences in format between the two methods. Such considerations include maintaining a neat personal appearance and developing a facility for displaying visual material to the respondent.

Prearranging the In-Person Interview. It is important to remember that in-person interviews must be prearranged in order to protect the privacy and safety of both the respondent and the interviewer, in contrast to telephone calling, which is performed spontaneously. In addition to refraining from making verbal reactions to the respondent, the interviewer should avoid any facial expressions or other gestures that may bias or otherwise disturb the respondent. It is recommended that all potential respondents be sent a letter not dissimilar from the one that introduces a mail-out questionnaire, including a description of the nature of the study and a statement concerning the importance of the recipient's participation. The letter should further state that a telephone call will soon follow in which the interviewer will seek to arrange an appointment for a personal interview at a place convenient to the respondent—often the respondent's home or place of work. Approximately one week after delivery of the letter, interviewers should begin placing the telephone calls. The guidelines for conducting these calls should follow the same format in terms of time of day and follow-up calling procedures as telephone interview calls.

Intercept Surveys

The intercept survey is a variation of the in-person survey whereby information can be obtained from respondents as they pass by a populated public area such as a retail mall, a workplace, a transit station, or an airport facility. The interviewer actually "intercepts" individuals and asks them to participate in the survey.

Advantages and Disadvantages. The advantages of the intercept survey can be stated as follows:

- *Complexity:* The availability of an interviewer provides the opportunity to explain unclear questions and use visual aids in the conduct of the interview. The use of maps and graphs is particularly important for transportation-related surveys.
- *Interviewer involvement:* The interviewer can ensure that questions are not skipped and that all questions that the respondent wishes to answer are completed.
- *Informs larger questionnaires:* The intercept survey is useful in informing the preparation of questionnaires for larger telephone, Web-based, mail-out, and in-person interview surveys and can also be helpful in structuring the discussion guide for focus groups.
- *Observed data:* The interviewer can observe certain personal characteristics of the respondent (such as gender, ethnicity, age, or physical disability), thereby avoiding the need to ask the respondent. This advantage serves to maintain the brevity of the survey and minimizes verbalizing potentially sensitive questions.
- *Cost-effective:* The intercept survey is more cost-effective than the traditional telephone, mail-out, and in-person surveys.

The disadvantages of the intercept survey are as follows:

- *Interviewer errors:* Interviewers may fail to follow a preestablished random procedure for selecting potential respondents. This can occur because interviewers may sense that certain potential respondents are threatening or unfriendly.
- *Limited information:* Since the intercept survey is of necessity quite short, the amount of information that can reasonably be obtained is limited. The researcher faces the challenge of asking the most important and relevant questions.
- *Lack of anonymity:* The anonymity of the respondent that is perceived in the telephone and mail-out surveys is greatly compromised in the intercept survey.
- *Interviewer bias:* The interviewer can inadvertently serve as a source of bias through hand gestures, body language, facial expressions, and extraneous comments; similarly, the respondent may respond to the interviewer in a less objective fashion in an effort to seek the interviewer's approval.

Administration of the Intercept Survey. The appropriate administration of an intercept survey requires adherence to certain rules and guidelines:

- A questionnaire should be prepared that takes about three minutes to administer. The survey should be pretested to ensure its effectiveness and brevity.
- The researcher should select a suitable location for the intercept survey. This location should be an area populated by potential respondents who are the target for the survey. The researcher should make sure that all legal requirements for the use of the location have been met.
- Ideally, a small incentive or prize should be offered to the respondent for completing the survey (for example, a pen or candy).
- Response rates are higher when the interviewer represents a public agency such as a city or county government since these affiliations tend to provide credibility to the research effort.
- Interviewers should wear a badge or some other form of observable identification that identifies them as part of the research project underway. The interviewer should carry legal identification in the event that the authorities challenge the interviewer's presence at the designated location.
- The researcher should establish time periods on weekdays as well as weekends as appropriate and necessary for the study to ensure that a reasonable cross section of the target population has been accessed.
- Interviewers, who reflect the demographic characteristics of the population to be surveyed, should be recruited and trained by the researchers. This training should include proper interviewing techniques, obtaining the attention and eventual participation of respondents, accurate recording of responses, and the importance of conviviality and confidentiality.
- The researcher must inform interviewers of the procedures to follow if they encounter problems during the administration of the survey instrument, procedures for contacting supervisors, and the importance of cleanliness and proper attire.
- The researcher should provide interviewers careful instructions concerning where to pick up blank surveys and where and when to return the completed ones.
- Survey forms should be prepared in the language or languages of the potential respondents, and interviewers must be selected who are bilingual in the appropriate languages. For intercept surveys in most states, survey forms must be prepared in both English and Spanish at a minimum. In several states, various Asian and African languages may also be necessary.

- The researcher should establish procedures for monitoring interviewers in the field as a form of quality control.

On-Board Surveys. The on-board bus and rail survey is a form of intercept that can be an effective and useful method for collecting information for transportation planning. It enables the researcher to gain information about travel patterns that aid in the planning of bus and rail routes. Also, it elicits opinions about customer satisfaction with bus and rail service so that service improvements that meet the needs of bus and rail riders can be considered in the planning process. The on-board survey involves placing an interviewer on board buses or rail cars to interview riders in person or to provide them with survey forms for them to complete and return on-board or by mail. Many of the instructions provided for the administration of the more general intercept survey can be readily adapted for the on-board surveys.

Stages of the Survey Research Process

To conduct any of the five major types of surveys in a rigorous and unbiased fashion, it is important to adhere to specific procedures and apply them in a systematic manner. Although the stages are presented here as distinct steps, there is actually a great deal of overlap as the survey research process is pursued and implemented. An overview of the process is presented here, with each stage being explained throughout various chapters of this book. The following list displays these stages, which are explained more fully below it:

Stage 1: Identifying the focus of the study and method of research

Stage 2: Determining the research schedule and budget

Stage 3: Establishing an information base

Stage 4: Determining the sampling frame

Stage 5: Determining the sample size and sample selection procedures

Stage 6: Designing the survey instrument

Stage 7: Pretesting the survey instrument

Stage 8: Selecting and training interviewers

Stage 9: Implementing the survey

Stage 10: Coding the completed questionnaires and computerizing the data

Stage 11: Analyzing the data and preparing the final report

Stage 1: Identifying the Focus of the Study and Method of Research

During the initial stage, the researcher must be satisfied that sample survey research is the most appropriate method of collecting the necessary information for the study under consideration among the other potential data-gathering techniques of secondary research, direct measurement, and observation. The factors that make sample survey research appropriate are as follows:

- Adequate secondary data is not available.
- There is a desire to generalize findings from a small subpopulation to a larger population.
- The target respondent population is accessible.
- The data to be obtained is of a personal, self-reported nature.

Once survey research has been determined to be the most appropriate research method, the researcher has two fundamental tasks to consider. First, the goals and objectives of the study should be elaborated and refined, and second, the researcher should identify the specific format for collecting the data (mail-out, telephone, Web-based, intercept, or in-person interview). The latter decision, in particular, will be greatly influenced by the complexity of the data to be obtained, the accessibility of the sample population, the budget available for the study, and the time constraints that have been imposed for completing the project.

Stage 2: Determining the Research Schedule and Budget

Once the parameters and objectives of the study have been identified, the researcher must establish a timetable for completing the survey research project. The timetable should be flexible enough to accommodate unforeseen delays and yet be capable of satisfying the needs of the research sponsor. In conjunction with this timetable, a detailed budget should be prepared. Insofar as budgetary and time considerations permeate and constrain each step of the survey research process, it is critical that this stage be carefully implemented.

Stage 3: Establishing an Information Base

Prior to the development of a survey instrument (questionnaire), it is necessary to gather information about the subject matter under investigation from interested parties and key individuals. Such individuals might be brought together in an informal group setting where relevant issues and problems can be freely discussed and debated. The goals and objectives of the research can be clearly defined, and the practical relevance of the proposed survey can be explained. For example, a research organization may have the objective of studying the travel behavior and travel preferences of economically disadvantaged residents in a major city in the United States. At the outset, it would be important to hold a focus group meeting, where representatives of social service organizations such as the county welfare agency, economically disadvantaged residents, and the researchers involved in conducting the study gather to exchange ideas and concerns. It is hoped that an open and frank discussion will reveal the type of survey information that would be helpful in outlining key issues and identifying relevant sectors of the population to be targeted in such a study. A detailed treatment of the focus group technique is presented in Chapter Four.

In some research endeavors, the subject matter is found to be new or vague, and as a result of this lack of general knowledge, it is not immediately feasible to devise a series of specific questions to be used in a formal survey process. In such situations it may be necessary to conduct, as a preliminary technique, some form of semi-structured direct observation of the population using professional observers who are trained to record information about the subject population in a systematic way. Such semi-structured research techniques have been successfully used in anthropological and sociological studies of geographical, economic, and behaviorally distinct subcultures. This base level of information may then be used to devise a questionnaire for the formal survey process. Without such preliminary information, the survey questions could prove to be peripheral or tangential to the goals of the research study. A thorough reconnaissance of information at this point is critical in terms of producing a focused and well-directed study. This chapter has already provided some background discussion regarding information collection. Further and more detailed treatment of this topic is found in Chapter Two.

Stage 4: Determining the Sampling Frame

The population that is identified for formal interviewing derives from applying the sampling frame for the survey research project. The researchers must be confident that the sample possesses the knowledge and information required to fulfill the requirements of the research project. That is, the sample represents the population from which they are supposed to be selected (called the *general population*).

After the general population, or *universe,* is defined in a conceptual sense, a list of identifiable and contactable members of this general population must be obtained. It is from this list that a sample of respondents will be drawn. This list is called the *working population.* The sequence of steps that moves the researcher from the general population to the sample is known as the *sampling frame.* For example, in a survey project concerning residential preferences and relocation tendencies, the general population may be defined as one that has demonstrated some mobility within a given metropolitan area. One way of operationalizing this concept of mobility is to obtain a list of residents who have recently moved. Local utility companies record changes of address whenever a new gas or electric hookup is requested. New hookups within a given period of time could easily identify a mobile population. Concepts related to identifying an appropriate population are discussed in Chapter Nine.

Stage 5: Determining the Sample Size and Sample Selection Procedures

The researcher must attempt to select a sample that is an approximate microcosm of the working population. Generally, given equally representative samples, larger samples yield a higher degree of accuracy than smaller samples. The researcher must weigh the desired degree of accuracy against the increased time and cost that a larger sample size entails. Once the overall sample size is determined, several alternative procedures must be considered for selecting a sample. Foremost among these procedures are simple random sampling, systematic random sampling, stratified random sampling, and cluster (multistage) sampling. The theoretical basis of sampling is discussed in Chapters Six and Seven, the criteria for determining sample size are described in Chapter Eight, and various sampling procedures are given in Chapter Nine.

Stage 6: Designing the Survey Instrument

The development of the survey instrument or questionnaire is a crucial component of the survey research process. At this stage, the researcher must devise a series of unbiased, well-structured questions that will systematically obtain the information identified in Stage 1. Developing the questionnaire can be an extremely detailed and time-consuming process. Decisions must be made concerning the wording of questions and the format depending on whether the survey is an in-person interview, intercept, mail-out, Web-based, or telephone survey. The number of fixed-answer and open-ended questions must be determined, and the element of time with respect to questionnaire length should be considered. The longer the questionnaire is, the greater are the variable costs associated with its implementation, such as interviewing time, computerization of data, and production and distribution costs. Furthermore, longer questionnaires tend to lead to lower response rates. The questionnaire must be easily understood and internally consistent and must lend itself to appropriate and meaningful data analysis. Questionnaire design is fully discussed in Chapters Two and Three.

Stage 7: Pretesting the Survey Instrument

After a draft questionnaire has been prepared and the researcher believes that the questions will obtain the information necessary to achieve the goals of the study, it is important to pretest the instrument under actual survey conditions. During the course of the pretest, poorly worded questions will be identified and the overall quality of the survey instrument refined. Based on the experience of the pretest, the questionnaire will be fine-tuned for use in the actual survey process. The pretest is discussed in Chapter Two.

Stage 8: Selecting and Training Interviewers

Telephone and in-person surveys require trained interviewers. These interviewers can be selected from the student ranks, they can be trained professionals, or they can be part-time, nonstudent interviewers. Researchers select interviewers according to the nature of the study and the characteristics of the sample respondents.

Prospective interviewers should be thoroughly trained by the researchers in the use of the questionnaire. It has been found that when interviewers have facility with the survey instrument, they are better able to generate and sustain respondents' interest in the survey. Interviewers should receive specific instructions on conducting their interviews and should be given guidelines for handling uncooperative respondents. Interviewer selection and training is described earlier in this chapter.

Stage 9: Implementing the Survey

The implementation of the survey instrument is a critical phase of the research process. Care must be taken that the established random sampling procedure is adhered to and that the timetable is strictly maintained. Ensuring the privacy and minimizing the inconvenience of potential respondents should be a major concern. In addition, a number of ethical standards must be followed by the researcher in the conduct of the survey research process. The administrative procedures in the conduct of survey research are discussed earlier in this chapter.

Stage 10: Coding the Completed Questionnaires and Computerizing the Data

The final questionnaire must be formatted in such a way that responses can be entered directly into the computer for data processing. (We discuss coding and formatting issues in Chapter Two.) Once the questionnaires have been returned, the very important process of "cleaning up" the forms begins. This is especially important for mail-out surveys, where no interviewer was present to make certain that instructions were followed. The clean-up entails making certain that the appropriate number of entries have been marked for each question, ensuring that there are no extraneous responses, and making sure that enough questions have been answered to validate the questionnaire. All open-ended answers must be categorized and coded on the form itself for ready computer entry. A variety of statistical software packages are available for this purpose.

Stage 11: Analyzing the Data and Preparing the Final Report

The recorded data input must be summarized, placed in tabular or graphical form, and prepared for statistical analyses that will shed light on the research issues at hand, using statistical significance tests, measures of central tendency, determinations of variability, and correlations among variables. These formal statistics and data summaries form the basis of the report that will be the culmination of the survey research process. Chapters Ten through Twelve elaborate on the essential statistical concepts involved in the analysis of survey data. Chapter Thirteen provides guidelines for the preparation of the final report.

Monitoring and Supervision of the Interview Process

For larger projects, a supervisor should be hired by the researcher and should be expected to work at least twenty hours per week, especially in the early stages of the interview process. With smaller projects, the researcher may also be able to serve as the supervisor, thereby eliminating the need to employ additional staff.

Telephone interviewing and the scheduling of personal interviews are best conducted from a centralized facility. This tends to produce higher response rates compared with interviews conducted or arranged privately from interviewers' homes or offices. It also affords the supervisor ample opportunities to directly monitor telephone conversations by listening to them. When such direct monitoring takes place, the respondent must be informed.

When telephone interviews and scheduling are conducted from private locations, the supervisor should randomly select at least 10 percent of the proposed sample and call these households to verify that contact has in fact taken place and to ascertain the respondents' degree of satisfaction with the conversation.

The supervisor should review the interviewers' work, be available for questions, and have frequent contact with the interviewers in the form of regular telephone or personal conferences. The supervisor should be prepared to reassign cases among interviewers if this is necessitated by such factors as language difficulties or varying completion rates. Production objectives should be established in terms of number of interviews to be completed in a given amount of time. It is the supervisor's responsibility to constantly monitor interviewer performance in terms of these objectives.

Conclusion

This chapter has introduced sample survey research as a useful technique for gathering information. The fundamental advantage of this kind of research is the ability to generalize about an entire population by drawing inferences based on data derived from a small portion of that population. Sample survey research can be applied to any facet of descriptive data, behavioral patterns, and attitudinal information about societal preferences and opinions. The chapter presents the stages of the survey research process and it emphasizes the types of surveys (mail-out, Web based, telephone, in-person, and intercept) as well as the procedures for their administration.

EXERCISES

1. Identify and briefly discuss the stages of the survey research process.
2. Discuss the advantages and disadvantages of the five major types of surveys: mail-out, telephone, Web-based, intercept, and in-person interviews.
3. Prepare a short summary of the principles associated with the selection, training, and supervision of interviewers.
4. Discuss the historical development of scientific survey research. Comment on the usefulness of this technique in permitting the researcher to make inferences about large populations by interviewing relatively few from the population.

CHAPTER TWO

DESIGNING EFFECTIVE QUESTIONNAIRES

Basic Guidelines

At the heart of survey research is the questionnaire development process. The key considerations in this process, including the placement of questions within the survey instrument and their format in terms of the method of implementation (telephone, mail-out, Web-based, intercept, or in-person interviews), form the basis of this chapter. The discussion of these issues takes place within the context of sample questions and exhibits derived from actual questionnaires and survey instruments that we have implemented in our professional sample survey projects. This chapter discusses certain general, macro questionnaire development considerations; Chapter Three delves more deeply into specific questions—or a micro-level examination of questionnaire development.

Be aware that no questionnaire can be regarded as ideal for soliciting all the information deemed necessary for a study. Most questionnaires have inherent advantages as well as inherent flaws. The researcher must use experience and professional judgment in constructing a series of questions that maximizes the advantages and minimizes the potential drawbacks. The guidelines detailed in this chapter recognize that there are a large number of considerations that the researcher must address in the process of questionnaire development. Sound questionnaire construction is a highly developed art form within the practice of scientific inquiry.

In the initial stages of the survey research process, it is important to determine the relevant issues that bear on the purpose of the research. Because social science research spans so many disciplines, it is impossible for any researcher to be fully knowledgeable in all the fields of study that might call on survey research services and skills. In addressing the complex multidisciplinary nature of survey research in the social sciences, the researcher can respond in two ways.

First, the principal investigator often seeks to construct a team of experts who jointly plan and implement the research study. This team represents both technical expertise and substantive knowledge of the political, socioeconomic, and cultural environment associated with the project. Second, with or without such a research team in place, and as a prelude to the development of survey questions, the investigators must gather preliminary information about issues of importance from interested parties and key individuals. These issues will derive in whole or in part from the three types of information elaborated on in Chapter One: descriptive, behavioral, and preferential. This preliminary information is best generated in a group setting where issues and problems of relevance to the study can be debated, discussed, and refined openly and constructively. Foremost among these preliminary information-gathering techniques is the focus group. The focus group is a semistructured discussion among individuals deemed to have some knowledge of or interest in the issues associated with the research study. Group participants are brought together in roundtable discussions run by a group leader or moderator. The discussion that ensues should contribute significantly to an understanding of the key substantive issues necessary for the development of the questionnaire. A full discussion of the use of focus groups in the survey research process is presented in Chapter Four.

At the conclusion of this preliminary information-gathering stage, the key issues that have emerged must be outlined and specified. This list of issues should be submitted to members of the discussion groups for clarification, confirmation, and perhaps further explanation. After this review, the researchers can prepare a draft questionnaire or survey instrument. If the research study has been commissioned by public agencies or private clients, as is frequently the case, the draft questionnaire should be reviewed by these parties for content and to ensure that the questions are consistent with the objectives of the study.

Once the researcher is satisfied with the draft questionnaire, the next step is to conduct a pretest. A *pretest* is a small-scale implementation of the draft questionnaire that assesses such critical factors as the following:

- *Questionnaire clarity:* Will respondents understand the questions? The researchers may find that certain ambiguities exist that confuse respondents. Are the response choices sufficiently clear to elicit the desired information?
- *Questionnaire comprehensiveness:* Are the questions and response choices sufficiently comprehensive to cover a reasonably complete range of alternatives? The

researchers may find that certain questions are irrelevant, incomplete, or redundant and that the stated questions do not generate all the important information required for the study.

- *Questionnaire acceptability:* Such potential problems as excessive questionnaire length or questions that are perceived to invade the privacy of the respondents, as well as those that may abridge ethical or moral standards, must be identified and addressed by the researchers.

The sample size for the pretest is generally in the range of twenty to forty respondents; however, for very large sample surveys, it is not uncommon for a pretest to contain a larger sample. The researcher is not really interested in statistical accuracy at this point; rather, interest centers on feedback concerning the overall quality of the questionnaire's construction. Accordingly, the researcher will select respondents from among the working population but need not be concerned about selecting them through a random sampling procedure (Chapter Nine) or in accordance with sample size requirements (Chapter Eight). Because statistical inferences are not the primary intent of the pretest, the researcher can be particularly sensitive to cost and time considerations—hence, the relatively small number of respondents. For example, a study that attempts to obtain information about teenagers might conduct a pretest using one or two high school classes. The members of the classes would very likely be individuals in the appropriate age category, and the classes could be surveyed quickly, conveniently, and efficiently. Clearly, not all teenagers are high school students; therefore, the high school classes would not necessarily represent the exact characteristics of the respondents in the final study. However, this degree of precision in the selection of pretest respondents is not required. The only requirement is that the pretest respondents bear a reasonable resemblance to the study's actual general population.

Following the pretest, the researchers must revise the questions as needed. They may want to perform a further pretest if these revisions are extensive. Otherwise, the final questionnaire can be drafted and prepared for implementation in an actual study.

Introducing the Study

A questionnaire is a conversation, and, like most other conversations, it builds on itself, commencing with an introduction. It is important to inform potential respondents about the purpose of the study in order to convey its importance and alleviate any concerns that potential respondents are likely to have. From the researcher's point of view, there is a need to convince potential respondents that their participation is useful to both the survey's sponsor or client and the respondents themselves. Any concerns that respondents may have regarding time and inconvenience, confidentiality, and safety should be allayed. The respondent must be assured that all answers are valuable—that there are no "correct" or "incorrect" responses.

An introductory statement should contain certain components. First, the organization or agency conducting the study should be mentioned (unless that information itself would be biasing). The introduction should state the relationship between the sponsoring institution and the potential respondent. A great deal of credibility can be gained for the study if the sponsor is a governmental body that in some way represents the respondent. An introduction that contains a reference similar to the following can be quite successful in establishing credibility: "The City of Chicago is conducting a survey of residents in order to assess community opinions about services provided by your local police department."

A general statement establishing the objectives and goals of the study and the significance of the results to the respondents themselves should follow the client reference. Potential respondents are more likely to participate when they perceive that the study's findings will have a direct impact on their well-being—for example: "It is the purpose of this study to identify those needs that the residents of the city feel should be addressed in order to maintain a peaceful and secure community."

The basis of sample selection should be made clear in order to make the respondent understand that there are no hidden agendas or undisclosed motivations behind the questionnaire. The characteristics the respondent possesses that led to his or her inclusion in the sample should be clearly delineated—for example: "Chicago is particularly interested in the opinions of new residents, and as such you have been selected at random from a list of new residents of the city."

The respondent must be assured that his or her participation is valued and that answers are neither correct nor incorrect. He or she must be assured that participation is strictly protected in terms of confidentiality—for example: "You should know that there are no right or wrong answers and that your responses will be treated confidentially. Survey results will in no way be traceable to individual respondents."

Because of the more personal nature of telephone, intercept, and in-person interviews, the interviewer should, as a matter of courtesy, identify himself or herself by name and obtain permission to proceed with the survey questions.

A telephone, intercept or in-person interview preamble might also include some estimate of the time required in order to complete the questionnaire. In the case of a mail-out questionnaire, the respondent should be able to judge this by direct observation of the instrument received in the mail.

A mail-out questionnaire should also include brief return mail instructions, such as, "Please drop your postage-paid, preaddressed response in the mail by June 15."

Exhibit 2.1 is an example of a mail-out or Web-based introduction that addresses the issues discussed above. Verify that the preamble contains the essential information. Exhibit 2.2 is an example of a telephone interview introduction. Once again, cross-check the highlighted issues against the example.

Because of the personal, physical presence of the researcher in face-to-face interviewing, Exhibit 2.2 can be revised into a somewhat less formal, more conversational format in this type of questionnaire administration. Exhibit 2.3 reflects these changes.

EXHIBIT 2.1. MAIL-OUT INTRODUCTION.

Dear Baytown Resident [*applicable respondent characteristic*]:

We need your help [*participative value*]! The City of Baytown [*organization identification/credibility*] is conducting a survey of all households in the city [*basis of sample selection*]. The information you provide will be useful in helping your City Council provide services and programs to meet the needs and wishes of the residents [*goals and objectives of study*].

Please take the time to complete the enclosed questionnaire. There are no correct or incorrect responses, only your much-needed opinions [*responses neither right nor wrong*]. This form contains an identification number that will be used for follow-up purposes only. All responses will be treated confidentially and will in no way be traceable to individual respondents [*confidentiality*] once the survey process has been concluded. Please drop your postage-paid, preaddressed envelope in the mail by June 24 [*return mail instructions*].

Thank you for your assistance. We care what you think [*participative value*].

Sincerely,

Jean M. Wilson
Mayor [*credibility*]

EXHIBIT 2.2. TELEPHONE INTRODUCTION.

Good evening (afternoon/morning). My name is Thomas Smith [*interviewer's name*]. The City of Flint [*organization identification/credibility*] is currently conducting a survey of Flint residents [*applicable respondent characteristic*] concerning the future development of library facilities for the city [*goals/objectives of study*].

Your household was selected at random [*basis of study selection*] to provide information and opinions regarding library facilities in the city of Flint.

We would greatly appreciate a small amount of your time [*time*] and your input on this important issue [*participative value*]. There are no correct or incorrect responses, so please feel free to express your opinions [*responses neither right nor wrong*]. Your responses will be treated confidentially and will in no way be traceable to you [*confidentiality*].

May I ask you a few short questions [*time/permission*]?

EXHIBIT 2.3. IN-PERSON INTRODUCTION.

Hello, my name is Janet Johnson [*interviewer name*]. The City of Flint [*organization identification/credibility*] is conducting a survey of its residents [*applicable respondent characteristic*] concerning the city's future development of library facilities [*goals/ objectives of study*].

Your household was randomly selected [*basis of sample selection*] to provide information and opinions about library facilities.

Would you be willing to answer a few short questions [*time/permission*] on this important issue [*participative value*]? Please feel free to express your opinions, because there are no correct or incorrect responses [*responses neither right nor wrong*].

The questionnaire form we complete today will not be marked in any way that would identify you [*confidentiality*].

Sequence of Questions

The order in which questions are presented can affect the overall study quite significantly. A poorly organized questionnaire can confuse respondents, bias their responses, and jeopardize the quality of the entire research effort. The following series of guidelines for sequencing questions has been created to enable the researcher to develop a well-ordered survey instrument.

Introductory Questions

The first questions should be related to the subject matter stated in the preamble but should be relatively easy to answer. Introductory questions should elicit a straightforward and uncomplicated opinion or derive basic factual—but not overly sensitive—information—again, as would a conversation. The main purpose of the early questions is to stimulate interest in continuing with the questionnaire without offending, threatening, confusing, or boring the respondent.

For a study involving quality of life among Native Americans who reside on reservations, we developed a questionnaire that began with the following questions:

1. To what tribe do you belong?
   1. _____ Pala   4. _____ San Pasqual
   2. _____ La Jolla   5. _____ Rincon
   3. _____ Pauma

2. How long have you and your family lived on the reservation?
   1. _____ Less than 1 year
   2. _____ 1 and under 5 years
   3. _____ 5 and under 10 years
   4. _____ 10 and under 20 years
   5. _____ 20 and under 30 years

       6. ____ 30 and under 50 years

       7. ____ 50 years or more

  3. Please indicate your general level of satisfaction with life on the reservation using the following scale:

       1. ____ Highly satisfied

       2. ____ Satisfied

       3. ____ Neither satisfied nor dissatisfied

       4. ____ Dissatisfied

       5. ____ Highly dissatisfied

The first two questions are of a basic, factual nature. The third question, although eliciting an opinion, is uncomplicated; however, it is germane to the key focus and sufficiently stimulating to secure the respondent's continued interest.

Sensitive Questions

Certain questions deal with sensitive issues, such as religious affiliation, ethnicity, sexual practices, income, and opinions regarding highly controversial ethical and moral dilemmas. It is recommended that these questions be placed late in the questionnaire, for two fundamental reasons.

First, if respondents react negatively to such questions and decide to terminate their participation in the questionnaire, the information obtained on all previous questions may still be usable in the overall survey results, because enough information may have been obtained to warrant acceptance of the interview as a completed survey with only a few questions remaining unanswered. Second, if rapport has been established between the interviewer and the respondent during the course of the survey process, there is an increased likelihood that the respondent will answer sensitive questions that come late in the questionnaire.

Related Questions

Questionnaires generally have a certain frame of reference, as indicated by their goals and objectives. Within this overall context, there are several categories of questions. For instance, the questionnaire soliciting opinions from Native Americans contained questions relating to housing characteristics, schools, public services, crime and police issues, economic development, employment issues, transportation, tribal decision making, recreation, shopping patterns, and socioeconomic data.

Proper questionnaire design dictates that related questions be placed together within the questionnaire so that the respondent can focus and concentrate on specific issues without distraction. In order to facilitate this, it is sometimes appropriate to separate categories of questions by providing a distinct heading that characterizes each section. In terms of police- and crime-related issues, the following sequence can be considered to be an acceptable one:

1. How would you describe the current relationship between the police and your community?
   1. ____ Good   2. ____ Fair   3. ____ Poor   4. ____ No opinion
2. During the past five years, do you feel that this relationship between the police and your community has:
   1. ____ improved   3. ____ worsened
   2. ____ remained about the same   4. ____ no opinion
3. In what ways could police officers improve their performance?

_____

If these same questions were to be commingled with questions from other categories, the resulting questionnaire would be much less likely to produce clear, well-formulated responses. You should be able to verify this by examining this less acceptable question order:

1. Do you or other members of your family participate in the tribal council's decision-making process?
   1. ____ yes   2. ____ no

2. Would you be interested in participating in a job training program?
   1. ____ yes   2. ____ no

3. In what ways could police officers improve their performance?

_____

While it is generally desirable to arrange questions pertaining to a particular subject in the same section of the questionnaire, it is also important to be cognizant of creating a patterned series of responses. Consecutive questions that tend to evoke reflexive responses, given without adequate thought, should be minimized.

Note that the sequence of questions in Exhibit 2.4, which is part of a commercial business survey, could well produce an automatic, unidirectional set of responses unless the respondent is sensitized to the subtle but important differences among the questions. This process of sensitizing will tend to minimize the risk of reflexive responses and is accomplished in this example by italicizing the essential distinctions.

EXHIBIT 2.4. SERIES OF QUESTIONS DEMONSTRATING SENSITIZING OF RESPONDENTS.

1. What types of additional businesses, if any, do you feel are needed in the City of Poway to help serve *your business needs?* (Please check no more than three types of businesses.)
   Types of Businesses
   _____ Food/market
   _____ Food/specialty store (bakery, deli, etc.)
   _____ Restaurant/dinner house
   _____ Restaurant/other (specify) _____
   _____ Retail/department store
   _____ Retail/specialty store
   _____ Professional
   _____ Services/supplies/equipment
   _____ Light industry
   _____ Other (specify) _____
   _____ Other (specify) _____
   _____ Other (specify) _____

2. What types of additional businesses, if any, do you feel are needed in the City of Poway to help serve the needs of *your employees?* (Please check no more than three types of businesses.)
   _____ Food/market
   _____ Food/specialty store (bakery, deli, etc.)
   _____ Restaurant/dinner house
   _____ Restaurant/other (specify) _____
   _____ Retail/department store
   _____ Retail/specialty store
   _____ Professional
   _____ Services/supplies/equipment
   _____ Other (specify) _____
   _____ Other (specify) _____
   _____ Other (specify) _____

3. What types of additional businesses, if any, do you feel are needed in the City of Poway to help serve the needs of *your customers?* (Please check no more than three types of businesses.)
   Types of Businesses
   _____ Food/market
   _____ Food/specialty store (bakery, deli, etc.)
   _____ Restaurant/dinner house
   _____ Restaurant/other (specify) _____
   _____ Retail/department store
   _____ Retail/specialty store
   _____ Professional
   _____ Services/supplies/equipment
   _____ Other (specify) _____
   _____ Other (specify) _____
   _____ Other (specify) _____

Alternative approaches to minimizing this risk of patterned responses may include the more frequent use of open-ended questions (without fixed alternative responses), questions that change the order of the fixed responses from question to question, or questions that vary substantially in terms of wording or length. The potential disadvantages of such tactics are that the respondent's thought focus may be disrupted or the respondent might become confused, thereby defeating the purpose of grouping these questions in the first place. Because several considerations must be balanced in the grouping of questions, the pretest becomes of paramount importance to identify the potential for inadvertently eliciting response patterns and to minimize any such impact on the study.

Logical Sequence

There is frequently a clear, logical order to a particular series of questions contained within the survey instrument. For instance, an appropriate time sequence should be followed. If questions are to be posed concerning an individual's employment or residence history, they should be structured in such a way that the respondent is asked to answer them in a sequential or temporal order—for instance, from the most recent to the least recent over a specified period of time:

Please indicate your places of residence during the past five years:

1. Current:

   _____

2. First prior residence:

   _____

3. Second prior residence:

   _____

Filter or Screening Questions

Other portions of the questionnaire might involve establishing the respondent's qualifications to answer subsequent questions. Through what are called *filter* or *screening* questions, as shown in Exhibit 2.5, the researcher can determine whether succeeding questions apply to the particular respondent. The first question requires that some respondents be screened out of certain subsequent questions. Only those who have participated in the city's recreational program are asked how they learned about the program. Both existing participants and nonparticipants, however, are asked about their intended use of a community pool and preferred payment programs, with a further screening out of questions pertaining to pool use for those respondents who have no intention of using the pool at all.

EXHIBIT 2.5. FILTER OR SCREENING QUESTIONS.

1. Have you or other household members participated in the recreation program offered by the City of Poway Community Services Department during the past 12 months?
   1. _____ Yes (Please continue with Question 2)
   2. _____ No (Please skip to Question 3)

2. If yes, how did you find out about the City of Poway Recreation Program? (Please check only one)
   1. _____ *Poway Today*
   2. _____ *Poway News Chieftain*
   3. _____ Community Services Department recreation brochure
   4. _____ Poway Unified School District flyers
   5. _____ Friend/family member
   6. _____ Other (specify) _____

3. A community swimming pool is being planned for Community Park at Bowron Road. If you and/or your family members plan to use this pool, which of the following payment methods would you most prefer? (Check one) If you and/or your family members do not plan to use the pool, please go on to Question 4.
   1. _____ Unlimited-use membership (Annual fee)
   2. _____ Purchase in advance a specified number of visits for discounted price
   3. _____ Pay each time you or your family members use the swimming pool
   4. _____ Do not intend to use the swimming pool. (If you have checked this response, please skip to Question 6.)

EXHIBIT 2.6. SCREENING USED TO DISQUALIFY RESPONDENTS.

1. Are you registered to vote in the City of San Diego?
   1. Yes . . . . _____ (CONTINUE)
   2. No . . . . _____ (DISQUALIFY)
   3. Not sure . . . . _____ (DISQUALIFY)
   4. Refused . . . . _____ (DISQUALIFY)

2. Did you vote in the 1986 elections for mayor or U.S. senator?
   1. Yes . . . . _____ (ASK QUESTION C)
   2. No . . . . _____ (DISQUALIFY)
   3. Not sure . . . . _____ (DISQUALIFY)

Under some circumstances, filter questions may be used to disqualify certain respondents from participating in the survey process at all. Exhibit 2.6 draws from a telephone questionnaire that was used in a survey of registered voters. It was the intent of the survey to query not all registered voters but only those who were likely to vote. For purposes of the survey, those who were most likely to vote were considered to be those who had voted for the mayor or U.S. senator in the previous year's election. The survey screened out entirely those who did not satisfy the appropriate preconditions by providing explicit instructions for the interviewer concerning disqualification.

Reliability Checks

On occasion, when a question is important or is particularly sensitive or controversial, the degree of truthfulness or thoughtfulness of the response may be in doubt. In such situations, it may be appropriate to include in the questionnaire a check of the respondent's consistency of response by asking virtually the same question in a somewhat different manner and at a different place within the survey instrument.

In a survey research project seeking to identify the demand for market rate housing in downtown San Diego, the following question was asked of respondents:

Please indicate the likelihood of your choosing to live in downtown San Diego.

1. _____ Very possible
2. _____ Somewhat possible
3. _____ Not very likely
4. _____ Highly unlikely

The researchers suspected that there might be a casual or less careful response pattern to this question, in which respondents might indicate their willingness to live downtown without giving the matter adequate thought. Therefore, later in the questionnaire, this question was posed:

When you consider the possibility of living in downtown San Diego, do you feel

1. _____ Excited
2. _____ Interested
3. _____ Indifferent
4. _____ Uncomfortable
5. _____ Frightened
6. _____ Other; please specify: _____

_____

In this study, in order for a respondent to be considered a "possible downtown resident," he or she had to choose the first or second response to both questions. Because any other combination might indicate a tentative or inconsistent willingness to consider downtown as a possible place to live, respondents with such answers were not considered strong candidates for downtown living. Without the benefit of this reliability check, respondents who were less likely to live downtown might well have been wrongly included with those who were more inclined to do so.

Question Format: Open-Ended or Closed-Ended

Most questions in a questionnaire have *closed-ended* response choices or categories. Such questions provide a fixed list of alternative responses and ask the respondent to select one or more of them as indicative of the best possible answer. In contrast, *open-ended* questions have no preexisting response categories and permit the respondent a great deal of latitude in responding to them.

Advantages of Closed-Ended Questions

There are several advantages to closed-ended questions. One is that the set of alternative answers is uniform and therefore facilitates comparisons among respondents. For purposes of data entry, this uniformity permits the direct transfer of data from the questionnaire to the computer without intermediate stages. The respondent's answers can be directed by a fixed list of alternatives, which limits extraneous and irrelevant responses. An example of a closed-ended question is found below:[1]

What is the highest level of formal education that you have achieved?

1. _____ Some high school or less
2. _____ High school graduate
3. _____ Some college
4. _____ Four-year-college graduate
5. _____ Postgraduate degree

If, instead, the question were open-ended, as shown below, the responses might not be quite so specific:

How much education do you have?

_____

Another advantage is that the fixed list of response possibilities tends to make the question clearer to the respondent. A respondent who may otherwise be uncertain about the question can be enlightened as to its intent by the answer categories. Furthermore, such categories may in fact remind the respondent of alternatives that otherwise would not have been considered or would have been forgotten.

Sensitive issues are frequently better addressed by asking questions with a preestablished, implicitly "acceptable" range of alternative answers rather than by asking someone to respond with specificity to an issue that might be considered particularly personal. For example, for medical purposes, an abortion clinic might require information about a client's history in terms of previous abortions. The questions, "Have you ever had an abortion? If so, how many have you had?" will tend to intimidate certain respondents who have had prior abortions and perceive that abortion carries a degree of social stigma. Their responses therefore might be biased toward minimizing the actual number. Recognizing that this tendency exists and always will in regard to socially sensitive issues, the researcher would improve response accuracy by constructing the question as follows:

How many abortions have you had?

1. _____ None
2. _____ One
3. _____ Two
4. _____ Three
5. _____ Four
6. _____ Five or more

Phrasing sensitive questions in this way, with alternative responses that extend significantly beyond normally expected behavior, implies that an accurate response is not outside the realm of social acceptability. (In this case, it implies that many other women may have similar histories and that having had an abortion is not necessarily aberrant behavior.)

Other types of sensitive questions may involve issues more closely associated with privacy than with social acceptability. This situation is encountered when the subject of a question is income. A respondent may very well feel that his or her privacy is violated when asked, "What is your annual household income?"

Giving alternative choices in the form of income ranges will tend to mitigate such feelings and will therefore generate a much higher level of response. A question about income is much better constructed to read as follows:

Please indicate the range that best describes your annual household income:

1. _____ Less than $15,000
2. _____ $15,000–$29,999
3. _____ $30,000–$44,999
4. _____ $45,000–$59,999
5. _____ $60,000 and above

Finally, fixed responses are less onerous to the respondent, who will find it easier to choose an appropriate response than to construct one. Thus, use of fixed-alternative questions increases the likelihood that the response rate for particular questions, and for the questionnaire in general, will be higher.

Disadvantages of Closed-Ended Questions

There are certain disadvantages to closed-ended questions that researchers should consider when developing a questionnaire. For example, there is always the possibility that the respondent is unsure of the best answer and may select one of the fixed responses randomly rather than in a thoughtful fashion. There is an increased possibility that the simplicity of the fixed-response format may lead to a greater probability of inadvertent errors in answering the questions. For instance, an interviewer or a respondent may carelessly check a response adjacent to the one that was actually intended. The advantage of ease of response therefore comes with some potential negative consequences. In a similar vein, a respondent who misunderstands the question may select an erroneous response.

Closed-ended questions in a sense compel respondents to choose a "closest representation" of their actual response in the form of a specific fixed answer. Subtle distinctions among respondents cannot be detected within the preestablished categories. This particular drawback is frequently addressed by inserting another alternative in the fixed-response format: "Other; please specify." This alternative represents an excellent compromise between closed- and open-ended response formats in that it is an open-ended question within a closed-ended format, as shown in this example:

Please indicate the activity in which you participate most frequently at the community recreation center.

1. _____ Basketball
2. _____ Volleyball
3. _____ Swimming
4. _____ Table games
5. _____ Aerobic exercise
6. _____ Other; please specify: _____

_____

For simplicity and ease of response, however, the use of this option must be carefully controlled. The decision to include an "Other" response category for a particular question must be based on evidence obtained during the pretest of the survey instrument. If the evidence shows that a relatively large number of responses to the question do not conform to the preliminary set of fixed alternatives (a minimum of 3 percent), then the researcher should formulate additional fixed categories for the responses that appear frequently and retain the "Other, please specify" category for the responses that appear less frequently. If there is no indication that an "Other" category is needed, it should not be included.

Using Open-Ended Questions

Researchers use open-ended questions in situations where the constraints of the closed-ended question outweigh the inconveniences of the open-ended question for both the researcher and the respondent. It is recommended that open-ended questions be used sparingly and only when needed. To the extent that they are used, the researcher must be aware of certain inherent problems.

First, open-ended questions inevitably elicit a certain amount of irrelevant and repetitious information. In addition, the satisfactory completion of an open-ended question requires a greater degree of communicative skills on the part of the respondent than is true for a closed-ended question. Accordingly, the researcher may find that these questions elicit responses that are difficult to understand and sometimes incoherent.

A third factor is that statistical analysis requires some degree of data standardization. This entails the interpretative, subjective, and time-consuming categorization of open-ended responses by the researchers. And finally, open-ended questions take more of the respondent's time. This inconvenience may engender a higher rate of refusal to complete the questionnaire.

Follow-Up Open-Ended Questions

As discussed, it is desirable to have relatively simple, fixed-answer questions wherever possible. However, most surveys find it necessary to seek information that cannot be fully answered within the fixed-answer format. In such cases, follow-up open-ended questions are asked in a manner that connects them to the fixed-answer question. For instance, during the studies of Native American tribes, the following questions were asked:

1. Are you generally in agreement with the policies and decisions made through tribal decision making?
    1. ＿＿＿ yes   2. ＿＿＿ no

2. If not, how do you generally differ?

_____

_____

Efforts should be made to place such open-ended questions as late in the appropriate section of the questionnaire as possible, while remaining cognizant of the need to have a logical and temporal order of questions.

Open-Ended Venting Questions

At the end of the entire questionnaire, it is often beneficial to use one or more open-ended "venting" questions—ones in which the respondent is asked to add any information, comments, or opinions that pertain to the subject matter of the questionnaire but have not been addressed in it. For example, a citizen opinion survey in a midsized suburban bedroom community posed the following final question in its questionnaire:

> *Thinking of your neighborhood as well as the city of Poway, in general, what do you personally feel are the most important issues or problems facing residents of this city?*

Questionnaire Length

The questionnaire should be as concise as possible while still covering the necessary range of subject matter required in the study. The researcher must be careful to resist the temptation of developing questions that may be interesting but are peripheral or extraneous to the primary focus of the research project.

The purpose of being sensitive to questionnaire length is to make certain that the questionnaire is not so long and cumbersome to the respondent that it engenders reluctance to complete the survey instrument, thereby jeopardizing the response rate.

As questions increase in complexity and difficulty, the questionnaire may be perceived as being tedious and longer than it actually is. Hence, the researcher must factor in such considerations as the number of questions and the time and effort required of the respondent to complete them.

As general guidelines, telephone interviews should occupy absolutely no more than twenty minutes of the respondent's time—and preferably closer to ten-to-twelve minutes; mailed questionnaires should take thirty minutes or less—preferably closer to fifteen minutes; Web-based surveys should also be targeted for fifteen minutes; in-person interviews should be limited to thirty minutes; and intercepts must be accomplished in four or five minutes at most—preferably three minutes.

Editing the Completed Questionnaire

An important part of the interviewer's task is to examine finished questionnaires for accuracy, legibility, and completeness. Despite this preliminary examination, the researcher must review each questionnaire for quality control purposes, especially with regard to filtering, multiple answers, and open-ended questions. Since mail-out questionnaires receive no intermediate interviewer examination, the researcher must be particularly careful in reviewing them.

In the review, the interviewer must be sure that questions that were designed to be skipped (through a filtering process) have indeed been skipped. If the interviewer has mistakenly asked an inapplicable question or has inadvertently marked a response to that question, the response should be deleted. In the case of questions that permit multiple responses and request a ranking, the first choice should be ranked by a code of 1 and the second choice by a code of 2. Such a question should be examined for accuracy in the following way:

- If only one response was made, it should receive a code of 1.
- Two responses should be coded with a 1 and a 2. If two responses are provided but are not ranked (they are indicated with a check mark, for instance), telephone or in-person interviewers should recontact the respondent immediately. This is an important reason for interviewers to examine the accuracy of their completed interviews at the time they are given. In the mail-out format, if there are only a few such responses, follow-up telephone calls, using the cross-referenced identification code, are in order. If there are many such inaccurately coded responses, the researcher can establish a new category for response categories that have been indicated but not ranked (see the discussion of postcoding below for the procedure for introducing new variable categories).
- More than two responses are not permitted. The telephone, intercept and in-person formats enable immediate corrections, and Web-based surveys can prohibit such errors as part of the programming. The mail-out can involve recontacting the respondent or creating an "Indicated but not ranked" category, and the final report should caution the reader that some respondents provided more than two responses.

After the review of the questionnaire has been completed, the researcher can begin the postcoding process.[2] In postcoding, responses to open-ended questions and "Other, please specify" response categories of the questionnaire are coded. To facilitate this process, the researcher should ask the interviewers to provide lists of all open-ended and "Other, please specify" responses.

With regard to "Other, please specify" responses, the researcher should first review these responses, identify those that reasonably belong to an existing, precoded category, and code them in accordance with that category by writing the code number directly on the questionnaire next to the response. This should be done in a different-colored ink from the one used to typeset the form and the one used by the respondent or interviewer to mark the questionnaire. This will permit the data entry technician to easily identify the postcoded response. The original "Other" response code should be crossed out for further clarity. "Other" responses that cannot be categorized into the precoded response categories can be treated in one of two ways, requiring a certain degree of judgment by the researcher:

1. When there is a sufficiently large number of the same or similar responses that do not belong in a precoded category, the researcher should consider creating a separate category with a new numerical code, starting with the first available number following the existing codes (the code for "Other" should also be adjusted to the highest code number so that it will print out last in the computer output). Therefore, it is not uncommon for "Other" categories to be precoded 9 if there are to be a few categories, or given a larger code number (for example, 20) if there might be a double-digit number of categories. If the frequency of any of these similar responses approaches the frequency of one of the existing categories, it is probable that a new code is warranted. This code should be marked on the questionnaire in a different color. Recoding is a frequent necessity in survey research in order to accommodate unexpected responses.
2. All responses that have a relatively low frequency of response can remain in a "Miscellaneous" or "Other" category.

The following question and answer can be used to demonstrate this process; a completed questionnaire contains a response that has been proven to occur with great frequency on other completed questionnaires and therefore merits a code of its own.

What kind of new business in Compton do you feel would give you the best opportunity for employment?

1. ____ Retail
2. ____ Light industry

3. _____ Heavy industry
4. _____ Office/professional
☒ _____ Other, please specify: *restaurant*

Open-ended questions require a similar postcoding process. That is, based on a verbatim listing of all responses to an open-ended question, the researcher again uses her or his judgment to develop categories into which these responses can be placed. The number of categories should be limited to approximately ten, with a maximum of fifteen to twenty, while adhering to the guideline that each should contain a respectable percentage (3 to 5 percent) of the total responses.

Table 2.1 was derived from the categorization of responses to the open-ended question, "How can the city government better serve your community?" By way of elaborating on the process of categorizing open-ended responses, the category of "Improve zoning/planning process" in Table 2.1 contains such verbatim responses as "fewer apartments," "more open space," "make developers pay fair share," and "protect property values."

EXERCISES

1. Choose a topic for a survey research study. Develop a list of at least five major interested institutions, organizations, or individuals whom you feel should be consulted for background information prior to the development of the questionnaire.
   a. What information would you seek from each of them?
   b. Whom would you select to pretest the draft questionnaire?

TABLE 2.1. WAYS IN WHICH CITY GOVERNMENT CAN SERVE COMMUNITY NEEDS.

	f	%
Provide improved local police protection	90	22.5
Ease traffic congestion	83	20.8
Enhance public education	74	18.5
Improve zoning/planning process	70	17.5
Provide more community funds	35	8.7
Improve communication	21	5.3
Other	27	6.7
Total	400	100.0

f = frequency or number of responses

2. What are the primary components to include in a preamble or introduction to a survey questionnaire? Write a preamble to a survey questionnaire that focuses on the demand and use of parks and recreational facilities in a medium-sized city.

3. Discuss the relative advantages and disadvantages of open-ended and closed-ended questions.

4. Define the following terms and concepts related to questionnaire development, and illustrate the concept with a question or series of questions, as necessary.
   - Sensitive questions
   - Related questions
   - Patterned response bias
   - Screening questions
   - Reliability checks

5. Write six questions for the parks and recreation questionnaire in Exercise 2. Include both open-ended and closed-ended questions, and place them in a sequence consistent with the principles outlined in the chapter. Identify the specific principles applied.

6. Use the following sample question to answer parts a and b below:

Which category best describes your occupation?

1. ____ Professional
2. ____ Clerical/Secretarial
3. ____ Sales
4. ____ Service
5. ____ Labor (other than construction and agriculture)
6. ____ Construction
7. ____ Agriculture
20. ____ Other (please specify)_____
_____

   a. Postcode the following "Other" responses, which occurred infrequently:
      - ____ Truck Driver
      - ____ Veterinarian
      - ____ Bookkeeper
      - ____ Cashier (Retail)
      - ____ Attorney
      - ____ City Manager
      - ____ Roofer
      - ____ Heavy Equipment Operator
      - ____ Gas Station Attendant
      - ____ Dancer
   b. Under "Other, please specify," a significant number of respondents answered "Retired." How would you postcode these responses?

7. Your staff has recorded and compiled open-ended responses to the following question into preliminary categories as designated in Table 2.2:

TABLE 2.2. IMPORTANT ISSUES FACING RESIDENTS OF THE CITY OF SAN ANTONIO.

	f	%
Road congestion	1,200	21.5
Too much growth and development	900	16.2
Need more open space	200	3.6
Preserve rural environment	100	1.8
Crime and drug abuse	350	6.3
School overcrowding	400	7.2
Poor-quality education	250	4.5
Parking problems	900	16.2
High taxes	175	3.2
Poor street maintenance	125	2.3
Need city beautification program	50	0.9
Inadequate library facilities	25	0.5
Inefficient government	150	2.7
Not enough jails	225	4.1
Need more jobs	300	5.4
Parks and recreation	200	3.6
Total	5,550	100.0

Thinking of your neighborhood as well as your city in general, what do you personally feel are the most important issues or problems facing the residents of this city? You may designate as many responses as you desire.

Using the principles and guidelines discussed in the chapter, establish the final categories for data entry (postcode) and recalculate the corresponding frequencies and percentages.

Notes

1. It is important that the categories of a closed-ended question be associated with numerical codes. The codes assigned during the questionnaire development stage are necessary for data entry and useful for data analysis.
2. When they have been completed, edited, and coded, the questionnaires are ready for the data entry process. There are myriad statistical software programs from which to choose. The researcher selects the most appropriate program based on the size and scope of the project, the sophistication of statistical analysis envisioned, the importance of the integration of graphics into the final report, ease of operation, and program cost. A major statistical program for survey research is the Statistical Package for the Social Sciences (SPSS). This program is sophisticated and comprehensive and capable of processing large amounts of data. It can generate both the very basic and the most highly advanced descriptive and analytical statistics and graphics. Microsoft Excel is also useful as a statistical analysis tool in the conduct of survey research.

CHAPTER THREE

DEVELOPING SURVEY QUESTIONS

The previous chapter addressed overall questionnaire development. No consideration of questionnaire development would be complete, however, without a thorough analysis of the principles and potential problems involved in the actual phrasing and formatting of the individual questions themselves.

Questionnaire construction is a skill that is refined over time by experience. Each research project has its own set of conditions and circumstances, which renders the imposition of fixed and rigid rules impossible. This chapter is particularly sensitive to the need for flexibility; instead of rules, it offers a series of objectives and guidelines in the pursuit of clear questions. Two fundamental considerations are involved:

- Question phrasing
- Question formatting

The researcher must use considerable discretion in the application of the guidelines outlined in this chapter, because there is a fine line between appropriately and inappropriately constructed questions. Such appropriateness can prove to be critical to the success of a research project.

Guidelines for Phrasing Questions

The way questions are worded is critically important to the success of a survey. Injudicious phrasing can lead to results that are ambiguous and potentially biased. The following guidelines are provided to assist in the preparation of survey questions that are objective and clearly worded.

Level of Wording

The researcher must be cognizant of the population to be surveyed in terms of the choice of words, colloquialisms, and jargon to be used in the questions. As a general guideline, wording should be simple, straightforward, and to the point. Specifically, the researcher should attempt to avoid highly technical words or phrases, words that require or are associated with higher levels of experience or education, and words or phrases that may be insensitive to ethnic- or gender-related issues.

For example, in a questionnaire seeking to obtain information related to the use of illegal drugs, the following possible questions might be asked:

- Have you or any member of your family been engaged in substance abuse during the past year?
- Have you or any members of your family used illegal drugs during the past year?

The first possibility uses the term *substance abuse,* which is not necessarily universally understood by the general population. Therefore, the responses to this question may not be consistent with its intent. The second possibility uses the simpler and clearer phrase *illegal drugs,* and the responses should consequently be more accurate.

Obviously, the researcher is interested in making certain that respondents understand the questions well enough to provide accurate representations of their opinions, behavior, and characteristics for purposes of the study. If questions are not understandable, one of three problems may arise:

- Information provided may be inaccurate.
- There may be a large number of "do not know" or "no opinion" responses.
- The rate of refusal to complete the questionnaire may be inordinately high.

Once again, the pretest looms large in importance in the detection of language-related problems.

On occasion, the general guideline of simplicity should be modified to accommodate special population groups. In a survey among attorneys concerning attitudes about courtroom procedures, it is appropriate to include words that are recognizable to those who have been formally trained in the law. If the survey were instead administered to the general public, the level of wording would of necessity be different.

Ambiguous Words and Phrases

Effort must be devoted to avoiding ambiguity in the questions. Ambiguity can occur from the use of vague words or phrases. For example, if a survey is seeking to determine the number of people residing together in one household, the question might be inappropriately worded, "How many people live in your household?"

It may not occur to respondents faced with this question that they should include themselves in the response. The confusion can be avoided by rewording the question in a clear and specific manner: "Including yourself, how many people live in your household?"

Similarly, in an attempt to determine household income, the question, "What is your income?" will produce a variety of unsatisfactory responses such as the respondent's annual income, the respondent's take-home pay, the respondent's hourly wage, or the total household income. What is generally sought in most surveys is total gross annual household income, before taxes. The question, "Please indicate the category that best represents your total annual household income, before taxes," will produce the desired responses.

Words such as *affiliate, identify, involved,* and *belong* often produce ambiguous results. For instance, asking an individual with which ethnic group he or she most closely identifies can be interpreted to mean, "With which group do I best get along?" rather than, "Of which ethnic group am I a member?" In the first interpretation, a respondent may provide more than one response in order to communicate a favorable inclination toward certain ethnic groups. However, the researcher is typically interested in ascertaining the respondent's own ethnic background and would find such a response uninformative. An appropriate phrasing for obtaining such information is, "Please indicate your race or ethnicity."

Another example of ambiguous wording is demonstrated in the following survey question: "Please indicate the number of organizations with which you are involved." The words *involved* and *organizations* are each sufficiently vague to be likely to generate a variety of interpretations among survey respondents. The specific organizational type (for example, social clubs, professional organizations) should be delineated, as should the precise nature of the involvement.

Confusingly Phrased Questions

Confusing questions generate a feeling of uncertainty in the respondent as to the intent or meaning of the question. If it is a closed-ended question, the respondent frequently does not understand which of the alternative responses will express his or her opinion on the issue. Consider the question in Example 3.1.

EXAMPLE 3.1

Does it seem possible or does it seem impossible to you that the Nazi extermination of the Jews never happened?

1. ____ It seems possible.
2. ____ It seems impossible.

The confusion can be eliminated with the following rewording of this question:

Does it seem possible to you that the Nazi extermination of the Jews never happened, or do you feel certain that it happened?

1. ____ It is possible that it never happened.
2. ____ I feel certain that it happened.

In the 1980s, many cities throughout the country decided to change the name of one of their major freeways or streets to reflect the memory of Martin Luther King Jr. One city that had changed the name of Market Street to Martin Luther King Way encountered substantial opposition from businesses on that street and decided to conduct a general survey among its registered voters to determine if there was support for the businesses' campaign to change the name of the street back to Market Street. One of the questions in this telephone survey was intended to sensitize respondents to the significance of Martin Luther King Jr. in U.S. political history in the hope that such sensitivity would prompt respondents to reject changing the name of the street back to Market Street. While only a portion of this question is reproduced as Example 3.2, the instructions to the interviewee (over the telephone) are illustrative of a confusing question. Another source of confusion can be found in instructions that are not explicit; for example, in relation to a question asking, "What are your favorite recreational activities from the list below?," the instructions might not inform the respondent how many activities to mark or whether these activities are to be ranked.

EXAMPLE 3.2

Using the numbers 1–5, with 5 meaning you're much more likely and 1 meaning you're much less likely, how likely would you be to vote in favor of changing the name of Martin Luther King Way back to Market Street if you learned that:

1. Blacks considered it an insult to change   1   2   3   4   5
   the name of the street back.
2. It meant that blacks would begin protesting   1   2   3   4   5
   and picketing.
3. The national news media saw it as an insult to 1   2   3   4   5
   blacks and gave the issue national attention.

Double-Barreled Questions

Double-barreled questions are those that might inadvertently confuse the respondent by introducing two or more issues with the expectation of a single response. An example of a multipurpose question might be, "Are you satisfied with traffic flow and parking availability in your community?" To respond to this question with a yes answer, the respondent would need to have a positive opinion of both traffic flow and parking availability, thereby denying the researcher potentially valuable information about each individual issue. Hence, such wording can result in findings for which the precise meaning is uncertain. Another example is found in a questionnaire that was published in a small-town newspaper in order to determine public opinion about future land development in the community. The first question in that survey was worded as in Example 3.3.

EXAMPLE 3.3

Do you believe the VISIBLE development at Alpine's freeway entrances will affect the image and property values of our whole community?

1. ____ Yes   2. ____ No

The only way to answer yes to such a question is to feel positively about both image and property values and about all of Alpine's freeway entrances. In other words, if a respondent considers such development satisfactory at one entrance and not at another or believes that image will be affected but not property values, a negative answer is appropriate. Hence, responses to such questions are impossible to interpret accurately. Any question that contains the conjunctions *and* or *or* should be reviewed very carefully for the possibility that it may actually be composed of more than one question.

Manipulative Information

Certain questions may require some form of explanation to be presented to the respondent in order to provide necessary background and perspective. The researcher must be very careful that explanatory statements do not unduly influence the response by providing biasing or manipulative information. The objective researcher should not skew responses in one direction or another, but rather should solicit genuine opinions, behaviors, and facts from the respondents. An example of such manipulation is as follows: "One of the Ten Commandments says, 'Thou shalt not kill.' Do you believe that the state has the right to exercise capital punishment?" More often, manipulative information is less obvious. The question in Example 3.4, adapted from a public opinion survey prior to a major local election, asked potential voters about funding for parks and recreation:

EXAMPLE 3.4

The federal government spends approximately $1,200 per U.S. resident on national defense. Do you believe that the federal government is adequately allocating funds for national parklands and recreational facilities by designating $10 per resident for this purpose?

1. ____ Yes   2. ____ No   3. ____ No opinion

Whereas the researcher may have provided the information about defense spending in order to provide perspective to the potential respondent, this information may also be manipulative by characterizing the funding for parks as comparatively inconsequential and therefore inadequate. A more straightforward question, without reference to the defense budget, may well generate an entirely different response. It is frequently the case that the researcher is interested, however, in how knowledge such as the difference between defense and national park spending might affect the respondent. That is accomplished by asking the straightforward question first and then following it up with the information, as shown in Example 3.5.

EXAMPLE 3.5

1. Do you believe that the federal government is adequately allocating funds for national parklands and recreational facilities by designating $10 per resident for this purpose?

1. ____ Yes   2. ____ No   3. ____ No opinion

1a. If you were to learn that the federal government spends approximately $1,200 per U.S. resident on national defense, would that change your opinion about the adequacy of allocating $10 per resident for this purpose?

1. ____ Yes   2. ____ No   3. ____ No opinion

Unfortunately, manipulative information is occasionally incorporated deliberately into a questionnaire. It is not uncommon for certain survey sponsors to want to use such surveys for publicity purposes or to influence voter opinion. The small-town newspaper survey referred to previously contains the question in Example 3.6, which can be considered to contain manipulative information.

EXAMPLE 3.6

Do you agree with the current Alpine Planning Group's recommendation to build public trails on public right-of-ways and, if needed for safety, to enlarge the public right-of-ways?

(This would enable our residents, as well as the outside public, to use Alpine-area public trails to access nearby Cleveland National Forest without crossing Alpiners' private property. This would also minimize the liability, insurance, privacy, and safety problems posed to property owners by allowing public access to private property.)

The manipulative information is in the lengthy explanation, which can serve to bias the respondent toward an affirmative response. The information contained in that explanation may or may not be correct. It is clearly subject to some interpretation. Furthermore, there may be a problem in invoking the endorsement of what might be perceived to be an organization or institution with particular expertise, as in the earlier reference to the Ten Commandments. Referring to the Alpine Planning Group does not present such a significant biasing problem, but the researcher must be cognizant of the biasing potential involved in citing authorities such as religious organizations or highly respected public figures.

This discussion should not be construed as indicating that all explanatory information related to a question is necessarily manipulative. An appropriate use of an explanatory statement is set out in Example 3.7.

EXAMPLE 3.7

In July 2007, the Department of Transportation will open a "high-occupancy vehicle" (HOV) lane on I-5 for carpools and buses. This will be a separate lane, from Via de la Valle to Palomar Airport Road, carrying traffic southbound in the morning and northbound in the afternoon. Use of the HOV lane will require that at least two persons be riding in the vehicle.

Will you use the HOV lane to commute to work or in the course of your work?

1. ____ Yes   2. ____ No

If a "Park & Ride" lot were available near the on-ramp, would you be more likely to use the HOV lane?

1. ____ Yes   2. ____ No

Inappropriate Emphasis

The use of boldfaced, italicized, capitalized, or underlined words or phrases within the context of a question may serve to place inappropriate emphasis on these words or phrases. However, emphasis can serve a constructive purpose when the researcher needs to clarify potentially confusing nuances that may exist within the questionnaire (see Exhibit 2.4) or when titles to published works are involved.

Devices for indicating emphasis are inappropriately used when they are designed to evoke an emotional response or to impose the researcher's concept of significance rather than leaving the determination of what is and is not important to the respondent. Such tactics tend to bias survey results. An example of inappropriate emphasis is found in Example 3.8.

EXAMPLE 3.8

The City of Saint Ignatius has long been acknowledged as *America's Finest City,* with glorious sunshine and warmth. Please rate Saint Ignatius as a place to work and raise a family.

Very Good	Good	Neutral	Poor	Very Poor
1	2	3	4	5
____	____	____	____	____

Note that this survey question is simultaneously an example of the improper use of an explanatory statement, resulting in manipulative information, and the improper use of italics, resulting in inappropriate emphasis. It also violates the double-barreled question guideline by asking the respondent to simultaneously rate Saint Ignatius both as a place to work and as a place to raise a family. Example 3.3 can also serve to illustrate inappropriate emphasis. Its focus on the word *visible* seems to be an effort to disturb the community's rural residents by ascribing some form of visual obtrusiveness to the planned development.

Emotional Words and Phrases

Although they may be clear, simple, and otherwise acceptable, certain words and phrases carry with them the power to elicit emotions. Survey questions must be as neutral as possible to obtain accurate results and to fulfill their obligation to solicit and welcome all points of view. Questions must invite true responses from the entire population and not induce the respondent into giving an answer other than the one he or she would normally give (see Example 3.9).

EXAMPLE 3.9

The cornerstone of our democracy, the Bill of Rights, guarantees freedom of speech. Do you believe that subversives have the right to advocate the illegal overthrow of the U.S. government?

1. ____ Yes   2. ____ No   3. ____ No opinion

The words *subversives* and *overthrow* evoke negative feelings in most readers. These words can lead the respondent to associate such negative feelings with a negative response to the question. The introductory sentence can be construed as manipulative information that also leads to the desired negative response.

In general, slanderous and prejudicial language must be avoided, as must language that conjures up specifically positive or negative images. The question, "Do you prefer mountain village–like commercial zoning instead of open car storage, industrial zoning at the entrances to Alpine?" heavily slants the respondent toward the commercial zoning choice through the use of the phrases "mountain village–like" and "open car storage" to modify the competing land use choices. Such a tactic is inappropriate in that zoning itself does not necessarily dictate design or ultimate use, and it is very possible to have an unattractive commercial development and an attractive industrial one.

Levels of Measurement

Survey data are organized in terms of variables. A *variable* is a specific characteristic of the population, such as age, sex, or political party preference. Each variable in a survey consists of a set of categories that describe the nature and type of variation associated with the characteristic. The variable *gender*, for example, consists of two categories: male and female. Certain opinions are solicited in terms of three categories of response: yes, no, and no opinion. Some variables, such as annual income, can have numerous categories of response, depending on the researcher's purpose and focus.

Variables used in a survey project have distinct measurement properties associated with their categories. These are referred to as *levels of measurement* or *measurement scales*. Some variables can only be classified into labeled categories (*nominal scale*); other variables can be labeled but are also intrinsically capable of being ranked or ordered (*ordinal scale*); and still other variables not only imply a labeling and ranking but also are associated with certain standard units of value that determine exactly by how much the categories of the variable differ (*interval scale*).

Nominal Scale

The nominal level of measurement involves only the process of identifying or labeling the observations that constitute the survey data. In the nominal scale, data can be placed into categories and counted only with regard to frequency of occurrence. No ordering or valuation is implied. For example, a variable such as "political party preference" might be categorized into three possible responses: Republican, Democrat, and Independent. These response categories only serve the function of enumerating the number of survey respondents who indicate their respective affiliations. No ranking or ordering of the parties is specified or implied. Similarly, no valuation unit is available to permit the determination of the extent of each respondent's affiliation.

Ordinal Scale

The ordinal level of measurement goes a step beyond the nominal scale; it seeks to rank categories of the variable in terms of the extent to which they possess the characteristic of the variable. The ordinal level of measurement provides information about the ordering of categories but does not indicate the magnitude of differences among these categories. An example of the ordinal scale can be found in the variable of education—specifically, with regard to highest academic degree received. Potential responses for this variable might include doctoral, master's, and bachelor's degrees or other formal education below the level of a bachelor's degree. It is clear that these categories possess an ordinality or ranking, but they do not by themselves reveal any specific measure of the amount of difference in educational attainment.

Interval Scale

The interval level of measurement yields the greatest amount of information about the variable. It labels, orders, and uses constant units of measurement to indicate the exact value of each category of response. Variables such as income, height, age, distance, and temperature are associated with established determinants of measure that provide precise indications of the value of each category and the differences among them. Whereas ordinal levels of measurement with regard to age, for example, might include categories such as infant, child, adolescent, and adult, interval levels of measurement for age would entail precise indications in terms of established measures, such as years, months, or days.

Variables that are naturally interval but are arranged in categories with a range of values (ages: Under 10, 10–19, 20–29, and so on) are technically ordinal because the categories are not capable of precise measurement without an adjustment that will be discussed in Chapter Five.

Formatting of Questions

Whereas open-ended questions are relatively easy to present within a questionnaire, requiring simply an ample number of lines for the respondent to write an answer in full, closed-ended questions entail a greater range of considerations. The major issues related to the layout of closed-ended questions make up the balance of this chapter.

Basic Response Category Format

In formatting response category alternatives, the primary guideline to which the researcher must adhere is clarity of presentation. The choices must be clearly delineated so as to provide no confusion to the respondent or to the researcher examining the responses. The researcher must be able to recognize precisely what response choice has been indicated. Of particular importance is that each question be unambiguously associated with one and only one response category, with no overlapping of categories. Generally, either a check box or a line is provided next to the responses, and the responses are organized with sufficient space between categories, usually vertical space but not necessarily if space is an issue.

There may be occasions when the researcher wishes to conserve space in order to keep entire questions and their associated response categories together on one page, for instance, or to save paper and printing costs. Questions that offer a relatively few response alternatives can be organized horizontally as long as adequate space is provided between the possible responses in order that the respondent can easily identify the appropriate place to indicate the response and not inadvertently mark the line on the wrong side of the answer. In Example 3.10, the horizontal structure is acceptable and conserves space.

EXAMPLE 3.10

In your opinion, does San Diego need a rail transit system?

1. ____ Yes   2. ____ No   3. ____ No opinion

Some questions ask the respondent to circle the appropriate response. We do not recommend this device, because circled responses tend to be less easy to read during the data entry process, in that they can overlap more than one response category if not carefully executed.

Number of Alternative Responses

As discussed in Chapter Two, it is important to have as comprehensive a list of alternative responses as possible within each closed-ended question. However, the researcher must be careful that the number of fixed alternatives does not become so unwieldy that it confuses or intimidates the respondent. Ideally, in a mail-out or Web-based survey, there should be fewer than ten response alternatives for each question. In some circumstances, it may be necessary to increase that number of responses to an approximate maximum of fifteen. If it is suspected (through professional judgment, previous knowledge, or the formal pretest) that there will be a large number of very distinct response alternatives to a question that will be somewhat difficult to combine and that those choices will each be represented by a respectable percentage (say, 3 to 5 percent) of the total responses, then the researcher is justified in expanding the number of alternative response categories to the maximum of fifteen. The balance of choices can be handled through the use of an "Other, please specify" category. When the number of alternative responses in an in-person or intercept survey is large, the interviewer can show the respondent a card with the choices elaborated on it. The maximum number of alternatives in such surveys can even be extended beyond fifteen, up to twenty. However, a lengthy response list becomes problematic in the telephone survey format, where fifteen to twenty response categories are far too many. The number must be held to a maximum of six for the respondent to be able to remember and choose among them as they are read aloud by the interviewer. In the alternative, if reducing the number of response alternatives is not in the best interest of the research, the telephone question can be asked in an open-ended fashion but with the likely closed-ended categories already precoded and input into the computer for direct entry without the need for substantial postcoding.

Structure of Ordinal Categories Containing Interval Scale Variables

Interval scale variables pose special problems for structuring the alternative ordinal response categories. By the nature of their scale, nominal variables and typical ordinal variables have clearly identifiable categories in which there is generally little latitude with regard to assigning cases. For instance, a survey planned for implementation at a local zoo contained a question designed to determine exhibit preferences among zoo visitors. It could be anticipated that such a question would elicit responses such as petting corrals, reptile exhibits, or a tiger pavilion. All responses could be placed in a few possible categories that would be both reasonable and informative. In contrast, a question concerning the age of a respondent has an infinite number of possible ranges and interval sizes into which responses can be categorized. If, for example, the respondent is forty-three years of age, category alternatives for this one answer alone might include "35–44," "40–49," "40–44," "38–50," "over 40," and "under 50." Hence, deciding on the structure of categories for interval scale variables involves a greater degree of judgment and discretion on the part of the researcher.

There are several guidelines and rules of thumb to consider in this decision:

- Ideally, interval scale categories should be as equal as possible in terms of their interval sizes. In the case of age, fixed intervals such as "0–9," "10–19," and "20–29" should be considered an appropriate starting point.
- Each category should contain a reasonable number of responses. A manageable number of categories should be provided. Categories with very few respondents should be avoided, and categories with a very large number of respondents might tend to obscure details that are important to the focus of the study.
- The boundaries of the categories should conform to traditional breaking points wherever possible. It is more desirable to use income categories such as "$10,000–$20,000" rather than "$11,100–$21,100."
- Each category should consist of responses that are evenly distributed throughout its range of values. This assumption is necessary in order to avoid a skewed distribution of responses and to facilitate statistical analysis. For example, suppose that a researcher is conducting a survey in which respondents must be graduates of a four-year college in order to participate. For the variable of age, the category of "20 and less than 25" should be avoided, because most college graduates are at least twenty-two years old. Hence, the anticipated distribution within the category would be skewed toward the upper ages rather than being evenly distributed. The pretest of the survey instrument is of particular importance in helping to predict whether these preestablished categories will yield a relatively even distribution.

- The categories must be mutually exclusive, so that every response has one and only one possible category.

It may not be possible to satisfy all of these guidelines in any given situation. A potential difficulty in the application of these guidelines occurs when traditional category boundaries conflict with the principle of nonoverlapping categories. In the case of income, for instance, categories with traditional boundaries such as "$30,000–$40,000" and "$40,000–$50,000" are not acceptable within the same question, because an individual who earns an annual income of $40,000 applies to more than one category. An acceptable alternative would be "$30,000–$39,999" and "$40,000–$49,999," which assumes that all responses are rounded to the nearest dollar (or "$30,000–$39,999.99" and "$40,000–$49,999.99" without that assumption). Observations that in theory can assume the value of any number in a continuous interval require class boundaries that are inclusive of all such possible values. The use of the terms *under* and *over* can obviate any problems in the assignment of observations to the appropriate categories in such continuous variables. In fact, it is equally valid to use this format for class boundary determination for all variables except those for which whole number values are the only possible responses (for instance, number of children in a household). Hence, an appropriate format for these income categories would be "$30,000 and under $40,000" and "$40,000 and under $50,000," because of its clarity and simplicity and its conformity with traditional class boundaries.

Another conflict in these guidelines might arise with regard to interval sizes. Although it is desirable to maintain equal interval sizes for an income distribution, this objective may not satisfy the guideline that each category of the variable receive a reasonable number of responses. Typically, the frequency of response declines at higher income levels. Therefore, researchers often expand the size of category intervals at the higher income ranges in order to ensure that a reasonable number of responses per category is maintained rather than burdening the report of survey findings with unnecessary detail that is of minor consequence to the study. There is an element of proportion that is also important in category construction. That is, the difference between annual incomes of $10,000 and $20,000 is effectively much more significant than the difference between incomes of $150,000 and $160,000. Furthermore, there will always be some individuals who earn enormous annual incomes. Intervals cannot reasonably be provided in anticipation of these relatively few responses. Therefore, income questions should provide an unbounded upper-income category to account for this likelihood. Age distributions and certain other socioeconomic variables also demonstrate these patterns of response and should be treated similarly. Example 3.11 shows a reasonable breakdown of income categories.

EXAMPLE 3.11

Please indicate the category that best represents your total annual household income.

    1. ____ Under $10,000
    2. ____ $10,000 and under $20,000
    3. ____ $20,000 and under $30,000
    4. ____ $30,000 and under $40,000
    5. ____ $40,000 and under $50,000
    6. ____ $50,000 and under $75,000
    7. ____ $75,000 and over

Generally the most important of these guidelines to maintain are mutual exclusivity and even distribution within the categories. Next in importance is a reasonable number of responses, with traditional boundaries and evenly configured categories least necessary to maintain when they conflict with other guidelines.

Order of Response Alternatives

The list of alternative responses may possess an inherent logical order. This order must be replicated in the elaboration of these categories within the question. Ordinal or interval data are obvious examples, as indicated in Example 3.12.

EXAMPLE 3.12

How would you rate your day at the Wild Animal Park?

    1. ____ Very good
    2. ____ Good
    3. ____ Fair
    4. ____ Poor
    5. ____ Very poor

It clearly would not make sense to reorder the responses in Examples 3.11 or 3.12. Nominal data categories, however, should be randomly listed so as deliberately to eliminate any potential biasing effects of a particular sequence. Therefore, when conducting telephone, intercept, or in-person interviews, the order in which these response choices are read to the respondent should be periodically shuffled. For budgetary reasons and computer coding purposes, this shuffling is frequently not feasible for mail-out and Web-based surveys. However, the sequence of response alternatives in mail-out and Web-based surveys is less of an issue because the respondent is able to review the choices more easily than in other interview formats.

Multiple Responses

On occasion, a question may require more than one response, as demonstrated in Examples 3.13 and 3.14. These two examples represent the two basic types of multiple-response questions: in the first, the respondent is asked to rank preferences; in the second, choices are indicated without regard to their order. In constructing the questionnaire, it should be made very clear to the respondent if more than one response is acceptable or if a ranking is requested.

EXAMPLE 3.13

What kinds of entertainment would you most like to have scheduled at the new Performing Arts Center? (Indicate your highest priority with a 1, your second priority with a 2, and your third priority with a 3.)

1. ____ Plays
2. ____ Musicals
3. ____ Lectures
4. ____ Classical music
5. ____ Rock music
6. ____ Country music
7. ____ "Popular" music
8. ____ Dance
9. ____ Other (please specify) _____

EXAMPLE 3.14

In what ways could police officers improve their performance? (Interviewer: If respondent indicates that no improvement is needed, check the first box.) Check the two most important.

1. ____ No improvement needed
2. ____ Concentrate on important duties such as serious crime
3. ____ Be more prompt, responsive, alert
4. ____ Be more courteous and improve their attitude toward community
5. ____ Be more qualified in terms of training
6. ____ Need more Native American policemen on the reservations
7. ____ Other (specify)_____
8. ____ Do not know

In questions where the researcher requests only one response but there may be an inclination on the part of the respondent to supply more than one, instructions to "check only one" must be very clear, as in Example 3.15.

EXAMPLE 3.15

For which of the following pool activities would you most prefer to have "adults only" time periods designated? (Check only one.) If you do not want designated "adults only" time periods, check the last choice.

1. ____ Lap swimming (exercise)
2. ____ Water aerobic exercise classes
3. ____ General recreational swimming
4. ____ Organized competitive swimming
5. ____ Instructional swimming (swimming lessons)
6. ____ Do not want "adults only" time periods

Scaled Responses

Some questions require the use of a scaled response mechanism, in which a continuum of response alternatives is provided for the respondent to consider. Example 3.16 demonstrates a Likert scale. A Likert scale entails a five-, seven-, or nine-point rating scale in which the attitude of the respondent is measured on a continuum from one extreme to another with an equal number of positive and negative response possibilities and one middle or neutral category.

The extremes of such scales must be labeled in order to orient the respondent. It is also acceptable to label each numerical category on the scale. Generally, scaled responses work best horizontally to allow respondents to perceive the continuum. Caution should be exercised to provide adequate spacing between alternatives in the layout of the question. Further, the instructions should not indicate only one direction for the answer. That is, instructions should not say, "Please indicate the degree to which you agree with the following . . ." Instructions must refer to agree and disagree so that the respondent is aware that either answer is acceptable. The correct phrasing is, "Please indicate the degree to which you agree or disagree with the following . . ."

EXAMPLE 3.16

Please indicate your level of agreement or disagreement with the following quotation: "Religion, not patriotism, is the last bastion of a scoundrel."

Strongly Agree				Strongly Disagree
1	2	3	4	5
____	____	____	____	____

The Likert scale works particularly well in the context of a series of questions that seek to elicit attitudinal information about one specific subject matter. Example 3.17

is an example of such a series of questions that seeks to elicit the attitudes of professional urban planners about their jobs and their degrees of satisfaction.

When a series of questions such as the one presented in Example 3.17 has the
same set of response categories, it would be prohibitively wasteful of space and
monotonous to list question after question for several pages. In such circumstances,
these questions can be efficiently grouped together in a matrix or gridlike format.

All scaled response series should adhere to certain principles:

- The number of questions in the series should generally consist of two to ten
  items, depending on the complexity of the subject matter and the anticipated
  tolerance of the potential respondents.
- The questions chosen for the series should cover as many relevant aspects of the
  subject matter under consideration as possible.
- The questions should be unidimensional; that is, they should be consistent and
  concerned substantially with one basic issue.
- The scale itself must be logical and consistent with a continuum.
- For each question in the series, the scale must measure the dimensions of response
  in the same order. For example, in Example 3.17, the high end of the scale always
  measures dissatisfaction, while the low end always measures satisfaction.

EXAMPLE 3.17

Please indicate your opinion concerning the following characteristics of your present job.

Characteristics of Present Job	(1) Strongly Agree	(2) Agree	(3) Neutral	(4) Disagree	(5) Strongly Disagree
Opportunity to gain increased responsibility					
Opportunity to influence internal agency policies					
Opportunity to grow professionally (enhance skills and abilities)					
Opportunity to provide a useful public service					
Recognition of my contribution to the agency					
Sufficient remuneration for my efforts					
Opportunity to develop congenial relationships among colleagues					
Adequate resources to perform any assigned tasks					
Adequate evaluation of the quality of my work					
Reason to take pride in my work					

Although the Likert scale is quite common in survey research, it is only one of several types of scales available to the researcher. For instance, when it is suspected that a great number of respondents will choose a middle response and the research requires that respondents choose between alternatives, an even-numbered scale with few choices will work well. Consider the question in Example 3.18 about police and community relations. The responses do not permit the respondent to simply state these relations are neutral; the question forces a positive or negative response.

EXAMPLE 3.18

Would you characterize the relationship between the police and your community as . . .

1. ____ Very positive?
2. ____ Somewhat positive?
3. ____ Somewhat negative?
4. ____ Very negative?

EXERCISES

1. Referring to the types of information that sample surveys solicit, as presented in Chapter One (descriptive, behavioral, and preferential), write two sample questions for each of these three informational categories. Verify that none of the questions violates any of the principles of question wording.
2. Identify the level of measurement for the following variables and their categories:
    a. Kinds of bears (polar, grizzly, black)
    b. Resort destinations (Puerto Vallarta, Miami Beach, Hawaii)
    c. Decibel readings at test site (under 100 dB, 100 to 200 dB)
    d. Army rank (general, colonel, sergeant)
    e. Income classification (upper, middle, lower)
    f. Cities in New Mexico (Santa Fe, Albuquerque, Truth-or-Consequences)
    g. Movie rating classifications (G, PG, PG-13, R)
    h. Richter scale seismic measurements (3.5, 6.0, 7.2)
    i. Distance to stars in light years
    j. Class rank of graduating seniors (first, tenth)
3. Consider the following questions from various sample surveys, and indicate what problem or problems exist in the question phrasing.
    a. Do you believe that undocumented immigrants should be allowed to receive TANF payments?

1. ____ Yes
2. ____ No
3. ____ No opinion

b. Do you believe the cultural arts are uplifting to the community?
   1. ____ Yes
   2. ____ No
   3. ____ No opinion

c. Please indicate the number of institutions with which you have a personal relationship.

d. Your city has one of the *most efficient* governments in the country. Please rate your city in terms of its overall efficiency.

1	2	3	4
Very favorable	Favorable	Neutral	Unfavorable

e. Are you satisfied with police and fire services in your neighborhood?
   1. ____ Yes
   2. ____ No
   3. ____ No opinion

f. Do you believe that euthanasia should be practiced if the patient is hopelessly ill or provides consent?
   1. ____ Yes
   2. ____ No
   3. ____ No opinion

g. Please indicate your income:
   1. ____ Under $40,000
   2. ____ $40,000 and under $60,000
   3. ____ $60,000–$100,000
   4. ____ $100,000–$125,000
   5. ____ Over $125,000–$200,000

h. What kinds of activities would you like to see more of in your community?
   1. ____ Plays
   2. ____ Music
   3. ____ Dancing
   4. ____ Lectures
   5. ____ Ball games
   6. ____ Movies

i. Indicate your two favorite sports by placing a *1* by your favorite and a *2* by your second favorite. If you only have one favorite, place a *1* by it and do not mark any with a *2*. If your two sports are equally favored, give each a *1* and do not mark a *2*. If you have no favorite, go to the next question.
   1. ____ Next question
   2. ____ Baseball
   3. ____ Football

4. _____ Basketball
5. _____ Ice hockey
6. _____ Other, please specify (do not include figure skating because it is not a sport) _____

j. Which of the following statements comes closest to representing your opinion about George W. Bush's determination to eliminate Saddam Hussein from Iraq?
1. _____ It's about time we had a President who respected law and order.
2. _____ This is as bad as the Crusades.
3. _____ I'm afraid that the world may see this as an abortion of national sovereignty.
4. _____ I approve.

4. [For students] Draft a survey instrument of approximately ten questions to be administered to other students in your program concerning their satisfaction levels regarding the curriculum and quality of instruction.

5. [For working professionals] Draft a survey instrument of approximately ten questions to be administered to personnel in your department concerning their overall job satisfaction.

CHAPTER FOUR

UTILIZING FOCUS GROUPS IN THE SURVEY RESEARCH PROCESS

Chapter Two introduced the concept of focus groups as an information-gathering technique in sample survey research. They are invaluable in this regard and are also used significantly in debriefing surveys. That is, it is possible that survey results may surprise researchers and that the reasons for results may not be entirely clear. Focus groups, therefore, can serve to help inform the results regarding issues that the questionnaire itself did not or could not address, along with their role in gathering information prior to the survey. Further, focus groups are valuable research tools for in-depth qualitative research apart from the sample survey process.

Focus group research is part of a broad category of research called *qualitative research*. Scientific sample survey research is *quantitative*—the distinction being that sample survey results are quantifiable to a known degree of accuracy because the sample is representative, whereas focus groups provide information without that known degree of accuracy because representativeness is not ensured (Chapter Nine discusses sample selection and representativeness in detail.)

Focus groups generally involve eight to twelve individuals who discuss a particular topic under the direction of a moderator. The moderator promotes interaction and ensures that the discussion remains on the topic. These discussions typically are conducted over a period of one to two hours. Focus groups are more formally known as *focused group depth interviews*. This more formal designation indicates several important characteristics of the focus group:

- The term *focused* implies that the discussion is a limited one that deals with a small number of fixed issues in a semistructured format.
- The term *group* indicates that individual participants share an interest in the subject matter of the discussion and that they will interact with one another during the course of the session.
- The term *depth* derives from the nature of the discussion, which is more penetrating and thorough than is possible in the sample survey research process.
- The term *interview* implies that a moderator directs and conducts the discussion and obtains information from the individuals in the group.

Inasmuch as sample survey research is the subject of this book and that qualitative research books exist that cover the topic thoroughly, this chapter's purpose is not to present every aspect of focus group research but to introduce the basics of the process and alert the survey researcher to the possibility of using focus groups in conjunction with a sample survey.

Uses of Focus Groups

Focus groups were described in Chapter Two as a useful way of securing information for purposes of informing the development of the questionnaire prior to its implementation. The focus group has many other research-based uses. The most prominent of these uses are the following:

- Deriving opinions and attitudes about products, services, policies, and institutions in the private and public sectors in order to identify consumer and user perceptions
- Obtaining background information about a subject in order to formulate specific research questions and hypotheses for subsequent use in more quantitatively oriented research techniques (such as sample surveys)
- Testing messages designed to influence or communicate with certain audiences (such as juries, consumers, or voters)
- Identifying creative and innovative ideas related to the subject of interest
- Interpreting and enriching previously obtained sample survey results

The earliest applications of focus group research were in the areas of audience response to radio broadcasts and the effectiveness of army training and morale films in the 1940s. Focus groups have since become significant contributors to research in public program evaluation, public policy development, public and private sector marketing, commercial and political advertising, communications, and litigation. Focus groups are qualitative in nature and do not represent scientifically drawn samples of the population (see Chapters Eight and Nine). Consequently, the results of focus group discussions cannot be used to make generalizations about a larger population with a known degree of accuracy, as can be done through an appropriately designed sample survey. Focus groups, however, are important tools. They can aid in gaining a deeper understanding of the subject matter prior to developing the final survey questions. They can further aid in analyzing underlying themes, patterns, and nuances that exist in the population but are difficult to delineate from the survey results.

There are four fundamental components to the focus group research process: planning the focus groups, recruiting the participants, implementing the discussion sessions, and analyzing the results. The first three components of the process are discussed in this chapter; however, inasmuch as this book is devoted primarily to sample survey research, the component involving focus group analysis is deferred to Chapter Thirteen, where it is presented in the context of how to interrelate the qualitative findings from focus groups with the quantitative findings from sample surveys.

Planning Focus Groups

The focus group planning process consists of several activities that must occur before recruitment of participants can take place. Foremost among these planning activities are the following:

- Identifying the critical characteristics of potential focus group participants
- Establishing the appropriate number of focus groups
- Choosing the most appropriate facility for conducting focus groups
- Determining the necessity for financial inducements to encourage participation
- Scheduling the focus groups at the optimal times of the day and days of the week

Identifying the Critical Characteristics of Participants

Fundamental to the focus group process is the establishment of the critical characteristics of the potential participants. As a general rule, focus groups are more effective when they consist of participants who share many of the same key characteristics. Homogeneous groups tend to exchange ideas and opinions more freely than do groups with widely divergent backgrounds. Participants in homogeneous groups have been found to relate to one another well, and they tend to generate a higher quality of input.

Therefore, the critical characteristics of the focus group participants must be identified early in the research process. For example, a local newspaper was interested in opinions from subscribers about possible changes in the types of features being considered for publication in various circulation areas. Key characteristics were identified as follows:

- Subscribers to the newspaper
- Residents of particular circulation areas
- Age of subscribers

Therefore, six focus groups were scheduled. Two focus groups were to be held in each of three circulation areas. The two focus groups in each area were to consist of newspaper subscribers—one consisting of subscribers between the ages of eighteen and thirty-five and the other consisting of subscribers over thirty-five years of age.

It is important to note that each critical characteristic can increase the number of focus groups significantly. For instance, if gender had been regarded as critical to the newspaper study, four focus groups would have been necessary in each of the circulation areas if absolute homogeneity were to be maintained. Frequently, however, budget and time constraints prohibit the luxury of maintaining strict homogeneity within each group. In a focus group study of transit use in a large urban county, for instance, the following characteristics were considered important to the composition of focus groups:

- Geographical location
- Ethnicity
- Transit users versus nonusers
- Age

In such cases, it is recommended that a maximum of two to three characteristics of the population be regarded as critical, with the other important characteristics being assigned a secondary status. In this example, three geographical areas of the county and four primary ethnic groups were identified for analysis, yielding twelve focus groups. These two characteristics (area and ethnicity) were selected because transportation research is highly location specific and because language and cultural issues are frequently paramount in effective intragroup communication. The secondary characteristics, transit use and age, were mixed into the twelve focus groups in sufficient number (at least two to three representatives of each secondary characteristic per focus group session) to evaluate those characteristics without establishing separate focus groups for them.

Establishing the Number of Focus Groups

As demonstrated above, the nature of the research, including the critical characteristics of the population, in large part dictates the appropriate number of focus groups to be conducted.

In general, the number of focus groups planned should be a minimum of two, with an upper limit in the range of ten to fourteen. At least two groups are necessary because the researcher must be certain that he or she is not simply observing a unique set of circumstances that may exist only among the participants of any one group. The upper limit of ten to fourteen groups is more flexible because of homogeneity considerations and because of the varying degree of detail that the researchers require. As a general rule, however, beyond ten to fourteen focus groups, information becomes very repetitive and new information is rare.

In this planning stage, it is advisable to determine a target number of focus groups at the outset, based on key population characteristics and research requirements. However, the researcher must retain sufficient flexibility to add more groups if the originally planned groups fail to produce meaningful results.

Choosing an Appropriate Facility

Focus group discussions should be conducted in an easily accessible, convenient location. Participants are much more likely to attend when the location is near their homes or work, is easily accessible, and has convenient parking that is either free or paid for by the focus group contractor. Focus group facilities typically include conference rooms in hotels, community centers, restaurant areas, and specially designed rooms with one-way glass for viewing by the research team from a private viewing room.

Each type of location has advantages and disadvantages, and the researcher must assess these advantages and disadvantages in terms of the characteristics of the focus group participants.

- *Hotel conference rooms:* Conference rooms are generally quite effective as focus group locations. They are particularly well suited to mainstream, middle-income populations. The rooms generally are comfortable, climate controlled, quiet, and equipped by the hotel to suit the needs of the group, often including audiovisual capability, the necessary tables and chairs, and refreshments.

On the negative side, such hotel accommodations can prove to be expensive and can be uncomfortable for groups such as disabled residents and certain low-income populations who have been found to feel somewhat out of place at such locations.

- *Community centers:* Community centers include senior centers, neighborhood recreation centers, libraries, and other specialized community facilities frequently geared toward a lower-income or otherwise disadvantaged population group, such as physically disabled persons and the elderly. These groups tend to feel more comfortable close to home and in a familiar environment. Often the center does not charge the focus group researcher for use of the facility or charges only a nominal fee. On the negative side, these facilities are noisier and less adequate in terms of equipment and tables and chairs. Refreshments must be purchased outside the facility and brought to it.

- *Restaurant meeting areas:* Restaurants are particularly effective for focus group sessions among lower-income groups at mealtimes. There are a number of ethnic groups, in particular, for whom mealtimes are important socially and are therefore too significant to surrender in order to attend a focus group discussion. In our own experience, the Latino and Vietnamese populations are especially amenable to this type of location, not only because of the social importance of mealtimes but also because of the feeling that their community and its local businesses are being supported when a local restaurant is used for the session. The restaurant location can frequently prove to be very economical for the researcher, but on the negative side, even with private banquet facilities, restaurants can be very noisy and distracting.

- *Specially designed focus group rooms:* Market research firms are interested in hearing comments from the public that are not influenced by the sponsor of the focus group. Therefore, these firms frequently use a specially designed room with a one-way glass behind which the researchers and their clients can listen to and view the discussion without being seen by the participants. Sometimes the participants are told that they are being viewed behind the glass, and sometimes they are not. Disadvantages of this type of facility are as follows: they are limited in number and hence not particularly convenient to a broad base of potential participants; the secretive nature of the one-way glass may violate certain standards concerning privacy and scientific research ethics; and when the one-way glass is disclosed to the participants, it loses a considerable amount of its original value to the researcher. Research sponsors, on the other hand, are particularly in favor of these facilities because they afford privacy and comfort, as well as allowing sponsors to relax, eat, and discuss issues among themselves.

Determining the Necessity for Financial Inducements

It is frequently necessary to provide a monetary incentive to potential focus group participants to secure their agreement to participate and encourage their ultimate attendance. When members of the general public are invited to discuss issues, such as transportation, that are important public affairs, we have found that an honorarium of at least forty dollars plus the availability of light refreshments (such as coffee, soda, cookies, cheese and crackers) is required in order to obtain a satisfactory rate of attendance at a one- to two-hour session. In certain cases, it is necessary to provide meals to focus group participants. When full meals are provided, honoraria are typically not necessary or can be reduced. In order to secure adequate attendance at private sector focus groups that test products and marketing messages, it is frequently necessary to offer honoraria of one hundred dollars.

When focus groups consist of community leaders, government officials, corporate executives, or other highly positioned persons, honoraria are not only unnecessary but can be regarded as inappropriate. Meals are acceptable for these leaders when the focus group sessions occur at mealtimes.

Scheduling the Focus Groups

The researcher should always try to conduct more than one focus group in a day in order to achieve certain economies of scale, including the use of one facility at a single rental charge and the use of personnel with a minimum of travel and downtime. Focus groups should not be conducted on weekends (Friday, Saturday, or Sunday) or on holidays, when it is difficult to secure attendance. Therefore, focus groups should be scheduled on Monday, Tuesday, Wednesday, or Thursday, with Monday the least preferable among these. In order to provide the opportunity for the working population to attend, at least one and frequently more of these sessions should be conducted outside normal working hours.

When it is expected that the focus group will consist largely of mainstream working individuals, it is advisable to hold one meeting toward the end of the work day, from 5:00 P.M. to approximately 6:30 P.M. This meeting can be attended by individuals after work, who can then proceed home for dinner. A second focus group meeting, held from 7:00 P.M. to approximately 8:30 P.M., allows other participants to go home after work, eat dinner, and then attend the discussion. For groups for whom the availability of dinner is an important consideration, the dinner focus group should be conducted from 6:00 P.M. to approximately 8:00 P.M., with the first 30 to 45 minutes devoted to the meal. The other meeting on that day, if scheduled at all, should be conducted in the mid- to late afternoon, the precise time depending on the characteristics of the population.

Certain groups may not find any of these times to be suitable—for instance, train commuters who must adhere to a strict schedule in order to return home at the end of the day. The researcher therefore will encounter situations that may require adapting these scheduling guidelines to the needs of the specific population groups, but for the most part, the 5:00 to 8:30 P.M. time frame satisfies most focus group research needs.

Recruiting Focus Group Participants

As in the case of sampling (see Chapter Nine), the researcher must use a list of potential participants from which the actual participants are selected. The characteristics of the desired focus group participants dictate how the researcher should proceed to obtain this list.

There is no overriding rule for obtaining or generating such lists, but it is important to note that in contrast to scientific sample research, the list need not be exhaustive; it must simply include members with the desired characteristics. That is, the selection of focus group participants need not conform to the formal principles of survey research. Rather, participants are selected at the convenience of the researcher, so long as they possess the required characteristics.

When the focus group is to be composed of members of the general population without regard to specific population characteristics, generic sources, such as the telephone directory, various commercial directories, and computerized address and telephone directories, are usually suitable. If the requirements concerning characteristics of focus group participants are more circumscribed, then the researcher must be more creative in assembling such lists.

For example, a researcher who is conducting focus groups composed of representatives of the freight movement industry in a particular region might follow these steps in formulating the list:

- Identify the major modes of freight transportation: air, rail, truck, and ship.
- Contact various trade organizations, such as truck associations, if such trade organizations exist.
- Consult the Yellow Pages of the telephone directory for freight companies in all modes.
- Contact governmental jurisdictions, such as the Interstate Commerce Commission, for information regarding the existence of lists of freight companies.
- Ask each party contacted to identify additional potential invitees.

Sometimes the researcher's client possesses lists of customers, clients, and interested parties. These lists can prove to be quite valuable for recruiting purposes.

Once the list is developed, the researcher should begin telephoning a representative cross-section of the people or organizations on the list. The purpose of this call is to invite the potential participants to the planned focus group meetings. The potential participants are informed of the purpose and sponsor of the focus group session; the date, time, and place that it will be held; and the financial incentive or meal and refreshments to be provided.

If the person accepts the invitation, the researcher e-mails, mails or faxes a confirmation to the invited guest, detailing the specifications of the meeting (see Exhibit 4.1). The invited guest is reminded of the meeting one to two days prior to the focus group session by a follow-up telephone call.

The researcher should overrecruit in order to account for the fact that some guests will not show up for the appointed session. Our experience is that approximately 20 to 35 percent of confirmed guests will not attend focus group sessions.

EXHIBIT 4.1. SAMPLE FOCUS GROUP CONFIRMATION LETTER.

October 30, 2005

Ms. Samantha Houston
Bowie County Central Labor Council
2740 S. Harbor Blvd.
Crockett, CA 95000

Dear Ms. Houston:

The Bowie County Transportation District (BCTD) is in the process of updating its long-range strategic transportation plan. Public input has always played a vital role in shaping this plan. Therefore, just as with the initial plan, BCTD is seeking public input on this update.

In the current phase of the public information process, BCTD is requesting input from business and community leaders. We appreciate, therefore, that you, or someone you delegate, will be able to attend the scheduled focus group meeting, as follows:

DATE: Wednesday, November 12, 2005

LOCATION: Hannibal Hotel
Main Street
Crockett, CA 95000
555–5555

TIME: 5:00 P.M.

Please contact the undersigned if you find that you are unable to attend. We look forward to working with you to improve transportation throughout Bowie County.

Sincerely,

Joseph Hardy

Hence, in order to satisfy the objective that eight to twelve persons participate in a discussion, for example, it is best to confirm approximately fifteen guests.

Implementing the Focus Group Sessions

The implementation of focus group sessions can be divided into two components: preparation for the session and the actual conduct of the session.

Preparing for the Focus Group

The focus group room should be equipped with a table, preferably linen-covered and rectangular. The group moderator should be positioned at the head of a rectangular table or otherwise positioned so as to be visible by all the participants. The moderator should have a nameplate placed in front of him or her for reference by the group. Water and cups or glasses should be available to everyone at the table, and coffee, tea, and other refreshments should be available at stations against the wall. Videotape equipment should be positioned as unobtrusively as possible while still providing full coverage of the session. If, in the judgment of the researchers, videotaping will interfere with the discussion in any way, it should not be used. No method of recording should be used if any participant objects to being recorded. Alternatively, the objecting participant can be excused, and then the focus group can be recorded.

Chairs should be placed against the wall, out of the primary sight line of the participants to the maximum extent possible. These chairs are to be occupied by members of the research team, who will observe and record notes on the session. The researchers' client may also wish to attend and can be seated with the researchers.

A greeting table is set up at the entrance to the room in order to compile a sign-in list and to distribute nameplates to the participants for placement in front of them at the focus group table. These nameplates allow the moderator and other participants to address one another by name during the session.

Conducting the Focus Group

The focus group session should begin with an introduction by the moderator, including reference to the use of recording equipment, if applicable. This introduction should have three segments:

- A welcoming statement
- A brief overview of the subject matter to be covered
- An explanation of the discussion rules

The moderator then asks each participant to introduce himself or herself and indicate particular personal attributes that are important to the discussion at hand (such as place of work, place of residence, use of particular goods and services). Following these introductions, the moderator can proceed with the first question. The first question should be designed to engage all of the focus group members. It should be a relatively simple yet substantive question that can be answered quickly and will quell any participants' uneasiness about speaking in front of a group of unknown people.

The moderator can then move ahead with the balance of the questions, which are designed to elicit the requisite information. With each question, responses from the group are invited and discussed among the group until the moderator determines that it is time to go forward. Following each question, the moderator should summarize the views expressed and obtain some consensus that the summary is valid.

The moderator may determine that certain answers require further elaboration and probe for this elaboration by asking follow-up questions. Some follow-up questions can be anticipated in advance and scripted for the moderator as part of the questionnaire; others require impromptu adjustments by the moderator. Exhibit 4.2 presents an example of the moderator's structured question format that demonstrates the guidelines discussed above. The exhibit is drawn from a focus group discussion among Latino community members concerning the quality of bus service in their community and the marketing of that service to their community.

At the conclusion of the focus group session, ninety minutes to two hours after it began, the moderator can briefly summarize the overall discussion, ask if there are any further comments on the subject, thank the group for their time and valuable input, and provide the participants with the cash incentive, if promised.

EXHIBIT 4.2. STRUCTURED FORMAT FOR
A FOCUS GROUP DISCUSSION.

Moderator Introduction: "My name is Mary Doe. I am going to lead the discussion we have planned for today. We are here to discuss the transit system in our county—in particular, how the transit system can be more responsive to needs of the county's Hispanic community. We are going to discuss what works well for your community, what problems may exist from your perspective, and what suggestions you might have for improving the system. We are also going to talk about how the Metropolitan Transit Agency, or MTA, can better inform your community of public transit options. MTA is seeking to reach out to the ethnic communities of the county, and your input will be very valuable."

The moderator asks each participant to introduce himself or herself and to further indicate his or her job or profession and community of residence.

The moderator then explains the question-and-answer format to the group. Each person will give his or her response in succession, until all responses have been heard. The group will then openly discuss any comments, suggestions, or ideas, until the moderator determines that the question has been thoroughly discussed. The moderator will attempt to determine a consensus or validation of the views expressed and will summarize the key points.

Questions:
1. For what purposes does the county's Hispanic population use public transit, predominantly?
2. In what ways does the county's transit system work well for the Hispanic population?
   Follow-up questions (if not addressed in open discussion):
   a. What bus routes are used most often by your community?
   b. On what days is the bus system used more often? At what times of day?
3. In what ways does the system fail to meet your community's needs?
   Follow-up questions (if not addressed in open discussion):
   a. Is more service needed on some existing routes? Where and when?
   b. Is the fare charged for transit services satisfactory?
   c. Is there anything you want to say about safety issues or travel time considerations?
4. Now let us assume that MTA addresses many of the transit needs of the Hispanic population as you have identified them. We know from prior research that the general public often has difficulty obtaining and understanding information published by MTA about the services it currently offers. Therefore, your input as to how MTA can best make the Hispanic community aware of any improved services will be very valuable to everyone concerned. For example, if MTA were to advertise its services on the radio in the form of public service announcements or advertisements, what radio stations would be best to utilize [VERY IMPORTANT TO OBTAIN SPECIFICS], and how could these ads catch the attention of Hispanic listeners?
   Follow-up questions (if not addressed in open discussion):
   a. Should these ads be in English or make use of Spanish? If Spanish, in full or in part?
   b. Is there any special terminology that would help in catching the attention of the public and getting people to seriously listen to the messages?
   c. How about music? Should the tone be serious or light? Any ideas in particular?
   d. Are there any commercials you can think of which have caught *your* attention? Which ones? What about them worked for you?
   e. Are there any particular themes or issues of significance in the Hispanic culture that could be important to be aware of?
5. How about the use of newspapers or other local publications? Is this a good way to reach the community? Which newspapers or other publications [VERY IMPORTANT TO OBTAIN SPECIFICS]? What sorts of design or phrasing do you think would "catch the eye" of the reader?

EXHIBIT 4.2. (*Continued*)

Follow-up questions (if not addressed in open discussion):
   a. Again, is there any special terminology that would be helpful?
   b. Are there any particular themes or issues of significance to be aware of?
6. What about the use of television? How could this be used beneficially to inform the Hispanic community?
   Follow-up questions (if not addressed in open discussion):
   a. Are there particular cable programs or TV stations that could be used [VERY IMPORTANT TO OBTAIN SPECIFICS]?
   b. Are there particular public service–oriented programs that are frequently watched?
   c. Are there any commercials you can think of that have caught your attention? Which ones? What about them worked for you?
   d. Are there any special visual effects that would be particularly appealing?
7. Do you think that billboards, flyers, and posters can be effective in informing the Hispanic population of the county? How? What would work well? Why?
8. Are there particular public destinations or events of significance in your communities that would be better than others for an outreach program?
9. Is there anything else you would like to add in terms of reaching out to your community or in terms of services offered?

Thank you very much for coming here this evening. We and MTA are most appreciative of your having taken the time to contribute to this study.

PART TWO

ENSURING SCIENTIFIC ACCURACY

CHAPTER FIVE

DESCRIPTIVE STATISTICS

Measures of Central Tendency and Dispersion

The balance of the book concerns the statistical basis for sampling and analyzing and presenting sample survey data. This chapter introduces the statistical concepts that are critical to the scientific sample survey process.

Measures of Central Tendency

It is often useful to include *measures of central tendency* in the frequency distribution table to augment the description of the data presented. Measures of central tendency are statistics that provide a summarizing number that characterizes what is "typical" or "average" for those data. The three foremost measures of central tendency are the *mode,* the *median,* and the *arithmetic mean.*

Mode

The mode is that category of the variable that occurs most frequently. It is the only measure of central tendency determinable from nominal scale variables, but it can also be used to describe the most common category of any ordinal or interval scale variable.

TABLE 5.1.   POLITICAL PARTY AFFILIATION.

Party	Frequency (*f*)	Percent (%)
Republican	200	50.0
Democrat	150	37.5
Independent	50	12.5
	400	100.0

In the nominal scale data presented in Table 5.1, there are 200 Republicans, 150 Democrats, and 50 Independents. The modal response for these data is Republican. The mode is not the frequency of the most common variable category (200); rather, it is the category itself (Republican). The mode therefore conveys to the reader the category that is most typical of the population surveyed.

In retail sales, for instance, the mode takes on a particularly useful role, focusing the retailer's attention on the most popular products for purposes of identifying high-volume sizes, colors, or other characteristics. In transportation planning, traffic congestion is often described in terms of two daily peak (modal) periods: the morning and evening rush hours.

Median

The median is the category of the variable that represents the center, or midpoint, of the data. One-half of the data will have values less than the median's value, and the other half will have values greater than the median. To determine the median, the data must be capable of being arranged in order, from low value to high, or vice versa. As such, the median is determinable only for variables that are on an interval or ordinal scale. The nominal level of measurement does not lend itself to the computation of the median because it does not provide the qualities of order or rank.

For example, in a class of eleven students in a graduate seminar, the final exam scores are as follows:

Student	*Exam Score*
A	70
B	79
C	95
D	88
E	53

Student	Exam Score
F	80
G	98
H	93
I	76
J	85
K	82

The median for this exam score data can be identified in three steps:

1. Rank the data, from either low value to high value or vice versa.
2. Determine the median case location at $(n + 1) \times 0.5$.
3. Identify the category of the data that corresponds to the median case location.

The process therefore is as follows:

1. Rank the scores. 53, 70, 76, 79, 80, 82, 85, 88, 93, 95, 98.
2. Identify the median case location. The location of the median case is found by adding 1 to the sample size (n) and multiplying that sum by 0.5. Hence, in this example, the median location is $(11 + 1) \times 0.5 = 6$, or the sixth case in the ranked distribution of scores.
3. Identify the median category. With a median location equal to 6, the median itself is found by identifying the corresponding test score value in the ranked distribution of scores. That score is 82.

Score	53,	70,	76,	79,	80,	82,	85,	88,	93,	95,	98
Rank	1	2	3	4	5	6	7	8	9	10	11

Now suppose that a twelfth student missed the exam, took a makeup test, and received a score of 50. The median score would be that score that corresponds to a location of $6.5 = (12 + 1) \times (0.5)$. The corresponding exam score value occurs midway between 82 and 80, at a score of 81.

Table 5.2 demonstrates data that have been obtained by sample survey in the form of ordinal scale categories.

The determination of the median for this grouped ordinal data entails the same three steps discussed above:

1. Ensure that the categories are ranked from high value to low value, or vice versa. (Note that the ranking in Table 5.2 is on a continuum from the highest value, "very good," to the lowest, "very poor.")

TABLE 5.2.   RATING OF LOCAL POLICE SERVICES.

Rating	Frequency (f)	Percent (%)
Very good	100	21.1
Good	175	36.8
Fair	100	21.1
Poor	50	10.5
Very poor	50	10.5
	475	100.0

2. Determine the median location. This is found at $(475 + 1)(0.5) = 238$.
3. Identify the category of the data that corresponds to the median location. In grouped data as in Table 5.2, the identification of the median category requires the application of the concept of cumulative frequency. *Cumulative frequencies* are summations of frequencies from the frequency distribution. Each stated cumulative frequency represents the number of cases included in a particular category plus those cases already accumulated. In the use of Table 5.2, the cumulative frequencies are:

Rating	Frequency	Cumulative Frequency
Very good	100	100
Good	175	275
Fair	100	375
Poor	50	425
Very poor	50	475
	475	

The median location (case 238) can be found in the "good" category, in which cases ranked 101 to 275 occur (cases 1 to 100 being in the "very good" category). The median value for these data therefore is the rating category "good."

This same procedure can be applied to frequency distributions of data measured on the interval scale. Table 5.3 depicts such data derived from a residential sample survey.

The reader can readily identify a median location of 300.5 ($[600 + 1] \times 0.5$) and median number of children equal to 1. (The category contains cases ranked 251 to 325.) In Table 5.3, the data were measured on the interval scale with single values of the variable: number of children. Data derived from sample surveys, however, are generally grouped into ordinal categories containing ranges of interval data rather than data identified by single-value categories.

TABLE 5.3. NUMBER OF CHILDREN PER HOUSEHOLD IN ST. AUGUSTINE, FLORIDA.

Number of Children	f	%
None	250	41.7
1	75	12.5
2	125	20.9
3	75	12.5
4	50	8.3
5	20	3.3
6	5	0.8
	600	100.0

TABLE 5.4. CONSULTANTS' HOURLY FEES PAID BY CITIES IN NEW YORK, 1996.

Hourly Fees	f	%
$50 and under $75	30	7.5
$75 and under $100	80	20.0
$100 and under $150	140	35.0
$150 and under $200	100	25.0
$200 to $300	50	12.5
Total	400	100.0

Consider the results of a sample survey of hourly consultant fees paid by cities in New York State during 1996. These results are presented in Table 5.4.

When ordinal categories of a variable comprise a range of interval scale values (Table 5.4), in contrast to a single value of the variable (Table 5.3), the calculation of the median entails the application of four steps rather than three:

1. Rank the data from low value to high value only.
2. Determine the median location, in the same manner specified previously: $(n + 1) \times 0.5$.
3. Identify the category that corresponds to the median location, in the same manner specified previously.
4. Estimate the value of the median within the range of values in the median category.

The procedure for estimating the median is shown in Worksheet 5.1 and as Equation 5.8 in the endnotes of the chapter.

WORKSHEET 5.1.   DETERMINATION OF MEDIAN.

1. Verify that the categories are arranged from low to high value.

2. Determine the median location = case 200.5.

Fee	f	cf
$50 and under $75	30	30
$75 and under $100	80	110
$100 and under $150	140	250
$150 and under $200	100	350
$200 to $300	50	400
Total	400	

3. Identify the category that corresponds to the median location. Case 200.5 is in the "$100 and under $150" category, in which cases 111 to 250 occur.

4. Because these data are on an interval scale, it is possible to refine the estimate of the median further to a specific test score. To achieve this, it must be assumed that the 140 cases within the "$100 and under $150" category are evenly distributed throughout the category. Since there are a total of 110 cases below the median category, 90.5 additional cases (200.5–110) must be included from the median category in order to reach case 200.5. This means that 64.64 percent (90.5/140) of the median category must be included. Under the assumption of even distribution (Chapter Three), the median equals the lower limit of the median category ($100) plus the appropriate percentage (64.64 percent) of the category width of $50 (which represents the difference between the upper and lower limits of the category: $150–$100 = $50). Thus, the median can be estimated to be equal to $132.32, as shown below.

Lower limit of median category	= $100
Width of median category	= $50
Percent of median category required to attain median location	= 64.64
$100 + ($50 [.6464]) = $132.32.	

The determination of category width in Worksheet 5.1 depends on whether the variable being presented is discrete or continuous. The width of a continuous variable category is found by subtracting the lower limit of the category from the upper limit, as was the case in Worksheet 5.1 ($150 − $100 = $50). The width of a discrete variable category is found by counting all of the discrete units within the variable category, including the two end points.

In statistical analysis, variables with an infinite number of values between the end points of the categories are considered to be continuous. Examples of continuous variables are such measures as distance, time, and weight. Discrete variables, by contrast, have a limited number of possible values for the variable. Examples of discrete variables might be the number of retail establishments per city, the number of children per household, or the number of pages in a book. For discrete variables, a category such as "25 to 30 retail establishments" has a category width of 6, with the values of 25, 26, 27, 28, 29, 30 included in the category.

An exception to these descriptions of continuous and discrete variables is money. This variable is traditionally treated as continuous despite the fact that all currencies possess minimum measures for counting purposes. Such minimum measures, in essence, prevent the existence of infinite values between the end points of a category. For example, in U.S. currency, the penny effectively serves as this minimum measure.

Arithmetic Mean

The measure of central tendency that the general public most commonly uses is the arithmetic mean. It is the mean ($\bar{x}$) that most people most refer to as the *average*. The mean is the mathematical center of the data. It takes into account not only location of the data above or below the center (as the median does) but also the relative distance of the data from that center. The mean is, in essence, a point of equilibrium at which the sum of all distances from data points above the mean to the mean exactly equals the sum of all distances from data points below the mean to the mean. The mean requires that data be measured on the interval scale because the data are not only to be ranked but also to be measured.

In its simplest form, the mean can be demonstrated by using the same eleven test scores from the median example. It is calculated by summing all scores and dividing the total by the number of scores involved.

$$\bar{x} = \frac{\sum x}{n} \qquad (5.1)$$

where

$\sum$ = summation of all observations

x = value of each observation

Scores = 98, 95, 93, 88, 85, 82, 80, 79, 76, 70, 53

$$\bar{x} = \frac{899}{11} = 81.73$$

Generally, however, sample survey research data are more likely to be encountered in the form presented in Table 5.3 than listed individually. That is, in survey research, large numbers of observations are processed and organized into categories through the use of frequency distributions. When frequency distributions of interval scale data take the form of single-value categories, as in Table 5.3, the mean is calculated as follows:

$$\bar{x} = \frac{\sum fx}{n} \qquad (5.2)$$

where

Σ = summation

f = frequency of each category

x = single value of the variable category

For the data in Table 5.3, Worksheet 5.2 demonstrates the calculation of the mean.

In survey research, interval data are more often collected into ordinal categories comprising a range of values rather than a single value. As was discussed in Chapter Three, these types of categories are frequently used in closed-ended survey questions, and although they are literally ordinal in measure, they can be converted easily to interval through the application of category midpoints. Table 5.5 depicts such categories in a frequency distribution of travel time to work for a sample survey administered in Asbury Park, New Jersey.

WORKSHEET 5.2. CALCULATION OF MEAN NUMBER OF CHILDREN, ST. AUGUSTINE, FLORIDA.

Number of Children	f	fx[a]
0	250	0
1	75	75
2	125	250
3	75	225
4	50	200
5	20	100
6	5	30
Total $\bar{x} = \dfrac{\Sigma fx}{n} = \dfrac{880}{600} = 1.47$ children	600	$\Sigma = 880$

[a] x = number of children indicated for each category.

TABLE 5.5. TRAVEL TIME TO WORK IN ASBURY PARK, NEW JERSEY.

Travel Time (in minutes)	f	%
Less than 5	40	8.0
5 and under 10	54	10.8
10 and under 15	90	18.0
15 and under 20	102	20.4
20 and under 30	86	17.2
30 and under 40	70	14.0
40 and under 60	33	6.6
60 to 90	25	5.0
Total	500	100.0

The calculation of the midpoint assumes that the data are distributed evenly throughout the various categories. For purposes of calculating the mean for interval data presented in a range of values, each category must be characterized by a specific number, which instead of x will be the midpoint of the category range. Each midpoint is multiplied by the frequency for the category. These products are summed, and the result is divided by the total frequency to obtain the mean:

$$\bar{x} = \frac{\Sigma\,fm}{n} \qquad\qquad (5.3)$$

where

f = frequency of each category
m = midpoint of each category

Worksheet 5.3 can be adapted from Table 5.5.

WORKSHEET 5.3.   CALCULATION OF MEAN TRAVEL TIME TO WORK IN ASBURY PARK, NEW JERSEY.

Travel Time (minutes)	m[a]	f	fm
Less than 5	2.5	40	100
5 and under 10	7.5	54	405
10 and under 15	12.5	90	1,125
15 and under 20	17.5	102	1,785
20 and under 30	25.0	86	2,150
30 and under 40	35.0	70	2,450
40 and under 60	50.0	33	1,650
60 to 90	75.0	25	1,875
		500	$\Sigma = 11,540$

$$\bar{x} = \frac{\Sigma\,fm}{n} = \frac{11,540}{500} = 23.08 \text{ minutes}$$

[a]The midpoint for a category of interval scale data, consisting of a range of values, can be calculated by adding the lowest possible value of a category to the highest possible value for the same category and dividing that sum by 2. In the case of the category "5 and under 10," for instance,

$$m = \frac{5 + 10}{2} = 7.5.$$

This method for determining the midpoint applies in precisely the same manner to all categories on the interval scale (including both continuous and discrete categories). Hence the midpoint for "60 to 90" is calculated in precisely the same manner as "10 and under 15."

Measures of Dispersion

Measures of central tendency yield only partial information about a variable. They summarize data by identifying an appropriate average or center for those data; however, by itself, this average can lead to an incomplete, and at times misleading or confusing, description of the data.

To demonstrate the need to supplement measures of central tendency with additional statistical measures, consider the following example. Two similar communities are engaged in studies to determine the physical fitness of the firefighters in their respective fire departments. Firefighters were asked to participate in a test of stamina; Table 5.6 shows the distribution of endurance times for Community A and Community B. Longer endurance times are indicative of a higher degree of physical fitness.

Each community exhibits similar central tendencies with regard to endurance time (note that each arithmetic mean $\bar{x} = 6$ minutes). However, these communities also demonstrate significant differences. Specifically, Community A demonstrates a more consistent pattern of physical fitness, and Community B shows a wider spread of endurance times. Consequently, the mean endurance time for Community A is more descriptive of its data than is the mean endurance time for Community B. Community B demonstrates more variability in endurance time or greater dispersion in stamina.

To demonstrate this critical concept of dispersion further, consider that San Francisco and Washington, D.C., experience approximately the same mean average daily temperature in a typical year. It should be clear, however, that

TABLE 5.6. FIRE DEPARTMENT PHYSICAL FITNESS ENDURANCE.

Endurance (in minutes)	Community A		Community B	
	f	%	f	%
1 and under 3	8	4.0	30	15.0
3 and under 5	32	16.0	40	20.0
5 and under 7	112	56.0	70	35.0
7 and under 9	48	24.0	38	19.0
9 and under 11	0	0.0	13	6.5
11 and under 16	0	0.0	8	4.0
16 to 20	0	0.0	1	0.5
	200	100.0	200	100.0
	$\bar{x} = 6$		$\bar{x} = 6$	
	Median = 6.08		Median = 5.87	
	Mode = 6		Mode = 6	

San Francisco's consistent, temperate climate is significantly different from that of Washington, D.C., with its very warm summers and colder winters. Once again, the measure of central tendency by itself is not sufficient without additional information concerning the extent of the variability around it.

These examples demonstrate the need for an additional statistic capable of measuring the degree of dispersion associated with the central tendency. Measures of central tendency and measures of dispersion constitute the fundamental elements of what is known as descriptive statistics because they describe and summarize vast amounts of data by the use of single statistical values.

Range

The most elementary statistical measure of dispersion is the *range:* the difference between the highest and lowest values in the data under study. The range is calculated by subtracting the lowest value from the highest. In the fire department example, Community A has an endurance time range of 8 minutes $(9-1)$, and Community B has a range of 19 minutes $(20-1)$.

Two points should be noted from this determination of ranges. First, there is now a measure of dispersion that shows the greater consistency of times in Community A than in Community B. Second, the example reflects a disadvantage of the range, in that it can be affected by extreme values. In Community B, only 11 percent of all endurance times exceed the maximum endurance time in Community A, yet the differential in ranges between 8 minutes (Community A) and 19 minutes (Community B) can convey a much greater dissimilarity between these two communities than actually exists. Therefore, the concept of the range can be modified by eliminating from the analysis some portion of the low and high ends of the database.

The underlying principle of eliminating extreme scores from analyses is well accepted. For instance, international figure skating competitions are scored by first eliminating the lowest and the highest scores assigned to a skater by a panel of judges. In statistical analysis, there are several measures of dispersion that treat extreme scores similarly. The most common of these measures are the *interquartile range* (the difference between the values of the 25th and 75th percentiles) and the *decile deviation* (the difference between the values of the 10th and 90th percentiles).[1]

Standard Deviation

The interquartile range and other such modifications of the range improve on it by reducing the influence of extreme values. However, for most statistical research, these modifications are too severe because they eliminate a considerable amount of data. A measure of dispersion is desired that does not eliminate any values yet is not overly influenced by extreme values. The standard deviation is that measure. Rather than eliminating values, the standard deviation weights all values of the variable by their frequency of occurrence, thereby including extreme values but tempering their mathematical importance.

The standard deviation represents a version of a mean distance from each value of the variable to the arithmetic mean. The more dispersed the data are, the greater is the standard deviation.

Since the arithmetic mean represents that value at which the sum of the deviations from all data points above the mean and all data points below the mean balance out to zero, it becomes necessary to square these deviations in order to convert all deviations to positive values and eliminate the offsetting effect of equal positive and negative values. The squared distances are then divided by n in order to obtain a mean distance. The result of this process is the variance (s^2):

$$s^2 = \frac{\sum (x - \overline{x})^2}{n} \tag{5.4}$$

To convert the squared units of the variance back to units consistent with the data set, the square root of the variance must be calculated. This calculation yields the standard deviation (s):

$$s = \sqrt{\frac{\sum (x - \overline{x})^2}{n}} \tag{5.5}$$

Equation 5.5 is applicable to data consisting of individual cases. For data presented in frequency distributions with variables consisting of single values only, as in Table 5.3, the equation for the standard deviation is adjusted:

$$s = \sqrt{\frac{\sum f(x - \overline{x})^2}{n}} \tag{5.6}$$

In the case of interval data presented in ordinal categories consisting of a range of values, m (the category midpoint) is substituted for x:

$$s = \sqrt{\frac{\sum f(m - \overline{x})^2}{n}} \tag{5.7}$$

It is important to note that for small samples (generally less than 60), $n - 1$ must be substituted for n in equations 5.5, 5.6, and 5.7.

For the example of the eleven exam scores presented earlier in this chapter, the standard deviation can be calculated by operationalizing equation 5.5, as shown in Worksheet 5.4.

Worksheet 5.5 demonstrates the use of equation 5.6 (single-value category frequency distributions) in the calculation of the standard deviation pertaining to the number of children in St. Augustine, Florida (Table 5.3).

In the example of fire department endurance times (Table 5.6), the standard deviations for Community A and Community B can be calculated as demonstrated in Worksheet 5.6.

As indicated by the ranges (Community A = 8; Community B = 19) and the interquartile ranges (Community A = 1.8; Community B = 3.56), the standard deviations (Community A = 1.5; Community B = 2.82) also demonstrate that Community B is characterized by less consistent endurance times for its firefighters than Community A.[2]

The utility of the standard derivation with regard to its integral role in the explanation of sampling theory is discussed in detail in Chapter Six. As a descriptive statistic, the utility of the standard deviation derives from its indication of how closely the mean represents its data set. The lower the standard

WORKSHEET 5.4.   CALCULATION OF STANDARD DEVIATION FOR INDIVIDUAL CASE DATA.

1. Exam scores (x) = 53, 70, 76, 79, 80, 82, 85, 88, 93, 95, 98
$\bar{x}$ = 81.73

2. $s = \sqrt{\dfrac{\Sigma(x - \bar{x})^2}{n - 1}}$ (*Note:* number of cases < 60)

3.

x	$(x - \bar{x})$	$(x - \bar{x})^2$
53	−28.73	825.41
70	−11.73	137.59
76	−5.73	32.83
79	−2.73	7.45
80	−1.73	2.99
82	0.27	0.07
85	3.27	10.69
88	6.27	39.31
93	11.27	127.01
95	13.27	176.09
98	16.27	264.71

$\Sigma = 1{,}624.15$

4. $s = \sqrt{\dfrac{1{,}624.15}{10}} = \sqrt{162.415} = 12.74$

WORKSHEET 5.5. CALCULATION OF STANDARD DEVIATION FOR A FREQUENCY DISTRIBUTION WITH SINGLE-VALUE CATEGORIES.

1. $s = \sqrt{\dfrac{\Sigma f(x - \bar{x})^2}{n}}$ (*Note:* number of cases $\geq$ 60)

$\bar{x} = 1.47$

2.

x	f	$(x - \bar{x})$	$(x - \bar{x})^2$	$f(x - \bar{x})^2$
0	250	−1.47	2.16	540.00
1	75	−0.47	0.22	16.50
2	125	0.53	0.28	35.00
3	75	1.53	2.34	175.50
4	50	2.53	6.40	320.00
5	20	3.53	12.46	249.20
6	5	4.53	18.92	94.60
	600			$\Sigma = 1430.80$

3. $s = \sqrt{\dfrac{1430.80}{600}} = \sqrt{2.385} = 1.54$

WORKSHEET 5.6. CALCULATION OF STANDARD DEVIATIONS FOR TABLE 5.6.

Community A ($\bar{x} = 6$)

x	f	m	$m - \bar{x}$	$(m - \bar{x})^2$	$f(m - \bar{x})^2$
1 and under 3	8	2	−4	16	128
3 and under 5	32	4	−2	4	128
5 and under 7	112	6	0	0	0
7 and under 9	48	8	2	4	192
9 and under 11	0	10	4	16	0
11 and under 16	0	13.5	7.5	56.25	0
16 and under 20	0	18	12	144	0
	$n = 200$				$\Sigma = 448$

$s = \sqrt{\dfrac{\Sigma f(m - \bar{x})^2}{n}}$

$s = \sqrt{\dfrac{448}{200}}$

$= 1.5$ minutes

Community B ($\bar{x} = 6$)

x	f	m	$m - \bar{x}$	$(m - \bar{x})^2$	$f(m - \bar{x})^2$
1 and under 3	30	2	−4	16	480
3 and under 5	40	4	−2	4	160
5 and under 7	70	6	0	0	0
7 and under 9	38	8	2	4	152
9 and under 11	13	10	4	16	208
11 and under 16	8	13.5	7.5	56.25	450
16 and under 20	1	18	12	144	144
	$n = 200$				$\Sigma = 1,594$

$s = \sqrt{\dfrac{\Sigma f(m - \bar{x})^2}{n}}$

$s = \sqrt{\dfrac{1,594}{200}}$

$= 2.82$ minutes

deviation is, the better the mean reflects its data. Conversely, the greater the standard deviation is, the more dispersed the data and the less representative the mean.

The Normal Distribution

Most features or characteristics (variables) of a population tend to be distributed in accordance with the commonly understood concept of the bell-shaped curve. For instance, most adult American men stand between 5 feet 6 inches and 6 feet 2 inches tall, with far fewer less than 5 feet 6 inches or more than 6 feet 2 inches. If the heights of all American men were recorded and their frequencies plotted, the distribution would most likely resemble the bell shape depicted in Figure 5.1, which is known statistically as the *normal distribution* or *normal curve*.

In the normal distribution, the mean is located at the exact center and peak of the curve, dividing the curve into two symmetrical halves, each the mirror image of the other. That is, the mean, median, and mode are equal to one another in a perfectly normal distribution of data. The normal curve is also asymptotic to the *x*-axis. In other words, in both directions, the curve moves closer and closer to the *x*-axis but in theory never touches it. The possibility therefore always exists that there are findings that might be made outside the realm of existing knowledge.

FIGURE 5.1. THE NORMAL CURVE.

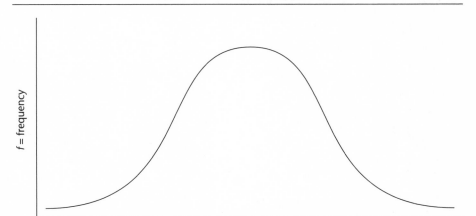

Most cases in the normal distribution are clustered around the mean. In the example of the heights of adult American men, if the mean height were 5 feet 10 inches, more men would be 5 feet 9 inches tall than 5 feet 6 inches. Similarly, more men would stand 5 feet 11 inches tall than 6 feet 0 inches.

Selecting the Most Appropriate Measure of Central Tendency for Describing Survey Data

When the three measures of central tendency are relatively close in value, it would be redundant to include more than one in a frequency distribution table. Moreover, when these measures differ in value substantially, it would be inappropriate and confusing to include more than one. Hence, the researcher should select only one measure of central tendency for inclusion in the frequency distribution table. This section outlines considerations that can aid the researcher in selecting the most appropriate measure of central tendency.

In the ultimate hypothetical normal distribution, the mode, median, and mean are equal. If, however, the distribution has a few extreme values, either high or low, the three measures of central tendency will begin to deviate from one another. Such a distribution is said to be *skewed*. Skewed distributions generally assume one of two forms:

1. The *positive skew* (Figure 5.2), which has a mean higher than the median and the mode because there are a few rather high values pulling the mean (and to a lesser extent, the median) toward these higher values

FIGURE 5.2. DISTRIBUTION WITH POSITIVE SKEW.

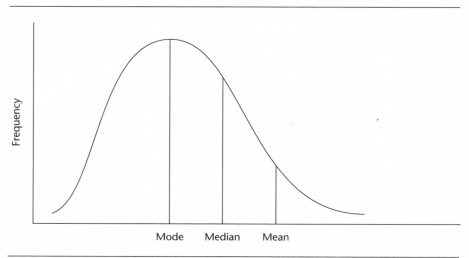

FIGURE 5.3. DISTRIBUTION WITH NEGATIVE SKEW.

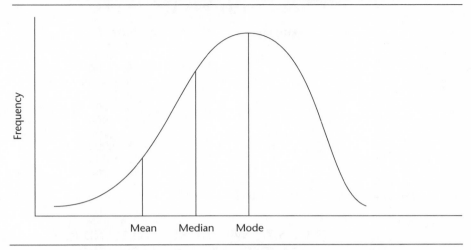

Mean   Median   Mode

2. The *negative skew* (Figure 5.3), which has some relatively low values pulling the mean and median toward the low values

In the case of a normal distribution, with the three measures of central tendency equal or approximately so, any one of these measures should suffice as a summary statistic for the data. In normal distributions, the rule for choosing the appropriate measure revolves around the level of measurement associated with the variable. Specifically, the mean should be used for normally distributed interval data, with the median being most appropriate for ordinal data and the mode remaining as the only measure of central tendency for nominal data.

The choice between median and mean under conditions of skewed interval data depends on how heavily the data are skewed. When the researcher determines that the mean is overly affected by the skew and therefore is a poor representation of the central tendency of the distribution, the median becomes the chosen statistic. Consider the following simple example.

The annual salaries of ten employees of a small private business are indicated in Exhibit 5.1. The mean income for these employees is $76,400 per year. It can be readily observed that this mean is not an appropriate representation of the central tendency of this income distribution, because nine employees out of the total of ten earn less than the mean income. A much more informative measure in this case is the median of $26,000. Skewness is often encountered with such variables as household income, age, and housing prices. The rule of thumb for such skewed variables therefore is that the best summary measure of central tendency, or average, is the median, not the mean.

EXHIBIT 5.1. SALARIES OF EMPLOYEES OF A SMALL, PRIVATE BUSINESS.

Employee	Salaries
A	$19,000
B	$21,000
C	$21,500
D	$22,500
E	$25,000
F	$27,000
G	$33,000
H	$40,000
I	$55,000
J	$500,000

Note: $\bar{x}$ = $76,400; median = $26,000.

TABLE 5.7. WILLINGNESS TO SPEND FOR EXPANDED PARKS AND RECREATION.

Annual Amount	Frequency
Nothing	37
More than $0 but less than $5	38
$5 and under $10	95
$10 and under $25	50
$25 and under $50	40
$50 and under $75	30
$75 to $100	25
Total	315

To explain this concept of skewness further, consider Table 5.7, in which the modal category is $5 and under $10 (often designated by the category's midpoint, 7.5), the median is $9.37, and the mean is $23.00. A positive skew is strongly suggested by these measures of central tendency, insofar as the mean is substantially higher than the median, which is in turn higher than the mode (see Figure 5.2). Such a skewed distribution is caused by a few respondents who commute long distances.

An indicator of skewness is the nature of dispersion, as measured particularly by the standard deviation. As discussed in Chapter Six, in a normal distribution, three standard deviations cover nearly the entire distribution. Therefore, the researcher can strongly suspect a skewed distribution if the mean, plus or minus three standard deviations, does not approximate the range of the data. In the case of the data in Table 5.7, note that the mean of $23.00 is associated with a standard deviation of \$26.37 and that $23.00 \times 3(\$26.37) = -\$56.11$ to 102.11; this is outside the range of the data on one side (the lower side) and roughly coincident on the other (higher side). The standard deviation will be sufficiently large to reach the farthest data extreme, and the result will be that one side will be substantially outside of the existing data range when the data are not normally distributed.

Finally, the most rudimentary way of determining whether a skew exists in a frequency distribution is to prepare a frequency polygon of a distribution.[3] In particularly skewed distributions, the frequency polygon will demonstrate the characteristics of skewness indicated in Figures 5.2 and 5.3.

Scaled Frequency Distributions

Another issue concerning the use of an appropriate measure of central tendency involves variables measured using a scale. The following is an example of a Likert scale question:

What is your general impression of how the Susanville city government affects your business?

Highly positive	Positive	Neutral	Negative	Highly negative
1	2	3	4	5
___	___	___	___	___

The responses to this question are shown in Table 5.8.

The reader may have noted that an arithmetic mean has been calculated for the data in Table 5.8, although the data are clearly ordinal in nature. As indicated above, the median, rather than the mean, is the appropriate measure of central tendency for ordinal data. In the case of Table 5.8, the median is found at case 300.5, which is contained within the category "Neutral." However, the scale, which

TABLE 5.8. IMPRESSION OF EFFECT OF SUSANVILLE CITY GOVERNMENT ON BUSINESS.

Impression	Value	f	%
Highly positive	1	100	16.7
Positive	2	175	29.2
Neutral	3	200	33.3
Negative	4	75	12.5
Highly negative	5	50	8.3
Total		600	100.0

Note: $\bar{x} = 2.67$.

generally associates numerical values with ordinal data (in this case, 1 to 5 for "Highly positive" to "Highly negative"), permits the researcher to calculate an arithmetic mean. In Table 5.8, this mean is calculated to be 2.67, which is slightly to the positive side of neutral. Hence, the mean has provided more information than the median. It has told the researcher that the center of the distribution is close to "Neutral" (the median) but is weighted somewhat toward positive response categories.

The calculation of the mean in this case makes important assumptions about the nature of the ordinal data by treating them as if they were interval data. First, it assumes that all respondents have a common understanding of the meaning of each response category. For example, in asking individuals how old they are, it is properly assumed that there is a common understanding of the concept of years in the measurement of age. Technically, this cannot be the case in terms of what respondents understand to be "Highly negative" versus "Negative," and so forth. Second, it assumes that there is an equal distance between each category of the variable that is measurable in accordance with the numerical values assigned to these categories. No such conclusion can be properly drawn with regard to ordinal data, inasmuch as ordinality affords the researcher the ability only to rank the data, not the ability to manipulate the data arithmetically. However, such manipulation has become accepted, because the power of the information obtained is considered to far outweigh the costs associated with relaxing these technicalities. Hence, it is recommended that in the case of scaled responses, the proper measure of central tendency should be considered to be the arithmetic mean, and in the case of a series of such responses, an arithmetic mean is an acceptable summary measure of the subject matter under study.

Example 3.17 (in Chapter Three) presented a scaled response series of questions administered to professional urban planners. The answers to those questions produced a mean score for each job characteristic, as follows:

Characteristic of Present Job	*Mean*
Opportunity to gain increased responsibility	2.81
Opportunity to influence internal agency policies	2.94
Opportunity to grow professionally (enhance skills and abilities)	3.19
Opportunity to provide a useful public service	3.10
Recognition of my contribution to the agency	3.17
Sufficient remuneration for my efforts	3.29
Opportunity to develop congenial relationships among colleagues	2.82
Adequate resources to perform any assigned tasks	2.85
Adequate evaluation of the quality of my work	3.18
Reason to take pride in my work	3.12
Overall mean	3.05

These responses, if described solely as ordinal data, would have yielded medians of "Neutral." However, the relaxation of a strict adherence to the ordinality of these data has allowed the researcher to identify important information not apparent in the use of the median. For instance, it is clear that the urban planners feel better about their opportunity to gain increased responsibility and their opportunity to relate to their fellow employees than they do about their compensation and their opportunity to grow professionally.

EXERCISES

1. The number of patients treated by a community clinic each day over a ten-day period was 52, 68, 39, 47, 57, 32, 75, 25, 31, and 93. Determine the arithmetic mean and the median number of patients for these ten days. Calculate the standard deviation.

2. A survey of apartment rentals is made in three counties. In County A, it is found that 230 rental apartments have a mean monthly rent of $820; in County B, 190 apartments have a mean monthly rent of $960; and in County C, the mean monthly rent for the 320 surveyed apartments is $640. What is the mean monthly rental for the 740 rental units in the survey?

3. A survey of 200 automobiles results in the following distribution:

Age of Auto in Years	f
0 and under 2	29
2 and under 4	43
4 and under 8	66
8 and under 15	38
15 and under 25	24
Total	200

   a. Calculate the mean age for all the autos examined.
   b. Calculate the median age for the autos.
   c. What is the modal class for these observations?
   d. Calculate the standard deviation.

4. Given the following distribution of income categories (in thousands), calculate the statistics indicated:

Income	f
$50 and under $60	8
$60 and under $70	6
$70 and under $80	12
$80 and under $90	9
$90 and under $100	5
Total	40

    a. Calculate the mean income.

    b. Calculate the median income.

    c. Calculate the standard deviation.

    d. (optional) Calculate the interquartile range.

    e. (optional) Calculate the decile deviation.

5. Consider the frequency distribution in Table 5.9.

    a. Calculate the mean and median for these data.

    b. (optional) Calculate the interquartile range and decile deviation.

6. Given the data from a sample survey of high school seniors shown in Exhibit 5.2, answer the questions that follow:

    a. Verify whether these data are basically characteristic of a normal distribution. ($\bar{x} = 973.67$; $s = 230.92$)

    b. Identify the most appropriate measure of central tendency for this distribution.

7. Consider Table 5.10, and answer the questions that follow.

    a. Identify the most appropriate measure of central tendency for the data in Table 5.10.

    b. (optional) Prepare a frequency polygon for the data in Table 5.10.

8. The following question was asked of citizens in a sample survey of a small city. The number in parentheses represents the number of respondents who indicated each category choice:

TABLE 5.9. HOURLY EARNINGS OF BLUE-COLLAR WORKERS.

Wage Rate	f	%
$5 and under $10	215	50.6
$10 and under $15	125	29.4
$15 to $25	85	20.0
Total	425	100.0

EXHIBIT 5.2. SAT SCORES: HIGH SCHOOL SENIORS.

SAT Scores	f	%
1400–1600	20	3.3
1200–1399	80	13.3
1000–1199	161	26.8
800–999	209	34.9
600–799	100	16.7
400–599	30	5.0
Total	600	100.0

TABLE 5.10.   ANNUAL HOUSEHOLD INCOME
FOR BENEDICT COUNTY FOR THE YEAR 1990.

Annual Income	f	%
Under $10,000	40	10.0
$10,000 and under $20,000	95	23.8
$20,000 and under $30,000	120	30.0
$30,000 and under $50,000	80	20.0
$50,000 and under $75,000	30	7.5
$75,000 and under $100,000	20	5.0
$100,000 and under $150,000	10	2.5
$150,000 and over	5	1.2
Total	400	100.0

In general, how would you characterize the effectiveness of your police department according to the following scale?

Highly satisfactory	Satisfactory	Neutral	Unsatisfactory	Highly unsatisfactory
1	2	3	4	5
(140)	(190)	(75)	(75)	(20)

    a.   Construct a frequency distribution table for this information.

    b.   Identify and calculate the most appropriate measure of central tendency.

9.   A home run hitting contest was held among amateur baseball players. Five hundred forty-five entrants hit a baseball as far as they could. Each was given three swings and only their best ball was counted. Table 5.11 indicates the best ball distance hit by the contestants.

    a.   Calculate the median long ball hit by these contestants.

    b.   What is the best way to explain what the median means?

    c.   Calculate the mean long ball.

    d.   Calculate the standard deviation.

    e.   Standard Deviation is a statistic that is called a measure of dispersion. What limitation to measures of central tendency do measures of dispersion seek to address?

10.   A survey of 315 residents of Smith County asked how much residents were willing to spend in taxes every year for an expanded parks and recreation system. The results are indicated in Table 5.12.

    a.   Calculate the mean amount that residents are willing to spend on parks and recreation.

    b.   Calculate the median

    c.   Calculate the standard deviation.

TABLE 5.11. DISTANCE TRAVELED BY BASEBALL.

Feet	Frequency
0 and under 200	50
200 and under 250	95
250 and under 300	110
300 and under 350	180
350 and under 400	155
400–450	95
Total	685

TABLE 5.12. WILLINGNESS TO SPEND FOR EXPANDED PARKS AND RECREATION.

Annual Amount ($)	Frequency
Nothing	37
More than 0 but less than 5	38
5 and under 10	95
10 and under 25	50
25 and under 50	40
50 and under 75	30
75–100	25
Total	315

Notes

1. Persons who prefer to calculate the median using an equation may use the following formula for this purpose:

$$\text{median} = \left[\frac{(n+1)(0.5) - \text{number of cases below median category}}{\text{number of cases in median category}} \right] \left(\begin{array}{c} \text{width of} \\ \text{median} \\ \text{category} \end{array} \right) + \left(\begin{array}{c} \text{lower limit} \\ \text{of median} \\ \text{category} \end{array} \right) \quad (5.8)$$

The median is also known as the 50th percentile, at which value 50 percent of the observations in the data set are greater in value and 50 percent are less. In general, it is possible to calculate any percentile (k) by determining the value of the variable at which k percent of total values are lower. The method for determining the kth percentile is a four-step process similar to the method used for the calculation of the median:

a. Ensure that the categories are ranked low value to high value.

b. Determine the location of the kth percentile:

   location of kth percentile = $(n + 1)$ $(k/100)$

   where k = percentile to be calculated.

c. Identify the category that corresponds to the kth percentile location, using cumulative frequencies.

d. Estimate the value of the kth percentile within the range of values in the *k*th percentile category.

Applying these steps to the data contained in Table 5.6 can generate the determination of the 75th percentile, for example. Worksheet 5.7 depicts the process of the 75th percentile calculation for Community A.

Persons who feel more comfortable calculating percentiles using an equation may use the following formula for this purpose:

$$k^{\text{th}} \text{ percentile} = \left[\frac{(n+1)(k/100) - \text{number of cases below } k^{\text{th}} \text{ percentile category}}{\text{number of cases in } k^{\text{th}} \text{ percentile category}} \right] \left(\begin{array}{c} \text{width of } k^{\text{th}} \\ \text{percentile} \\ \text{category} \end{array} \right) + \left(\begin{array}{c} \text{lower limit of} \\ k^{\text{th}} \text{ percentile} \\ \text{category} \end{array} \right)$$

(5.9)

2. There are derivations of these equations for the standard deviation that are easier to apply, but they are less intuitively clear. For ease of calculation, they are presented here, as follows:
   a. For individual case data:

$$s = \sqrt{\frac{\sum x^2}{n} - \bar{x}^2}$$

(5.10)

WORKSHEET 5.7. CALCULATION OF 75TH PERCENTILE ENDURANCE TIMES FOR COMMUNITY A (TABLE 5.6).

a. Categories ordered low to high

b. Location of 75th percentile

$$(200 + 1)\frac{25}{100} = 150.75$$

c. Identify the 75th percentile category

Endurance Times (in minutes)	*f*	*cf*
1 and under 3	8	8
3 and under 5	32	40
5 and under 7	112	152
7 and under 9	48	200
9 and under 11	0	200
11 and under 16	0	200
16 and under 20	200	

75th percentile category: 5 and under 7

d. 75th percentile: Since there are a total of 40 cases with endurance times below the 75th percentile category, 110.75 (150.75−40) additional cases must be included from the 112 cases in the 75th percentile category. This requires that 98.90 percent (110.75÷112) of the 75th percentile category must be included. Under the assumption of even distribution of values within categories, the 75th percentile itself equals the lower bound of the class (5) plus .9890 of the category width (2) = 6.98 minutes.

b. For frequency distribution with single value variables:

$$s = \sqrt{\frac{\sum fx^2}{n} - \bar{x}^2}$$

(5.11)

c. For categories with ranges of values:

$$s = \sqrt{\frac{\sum fm^2}{n} - \bar{x}^2}$$

(5.12)

Note, however, that these equations are applicable only in the event of samples consisting of sixty or more cases.

3. A graphic technique used for the display of interval data is the frequency polygon or line graph. A frequency polygon is constructed in the following manner:
    *   Locate the midpoints of each class interval along the horizontal axis.
    *   Plot the frequency associated with each class interval above the corresponding midpoint.
    *   Connect the points with lines.

Figure 5.4 is the frequency polygon corresponding to the data presented in Table 5.6.

FIGURE 5.4.   FREQUENCY POLYGON SHOWING FIRE DEPARTMENT EMERGENCY RESPONSE TIME IN COMMUNITY A AND COMMUNITY B.

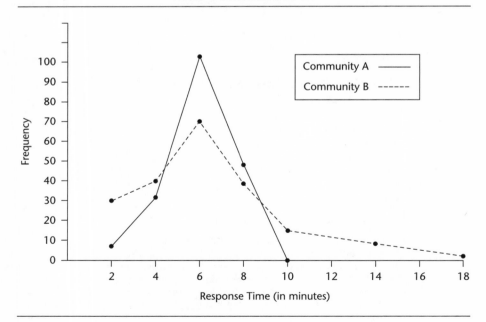

CHAPTER SIX

THE THEORETICAL BASIS OF SAMPLING

This chapter provides an introductory treatment of the statistical concepts that underlie the essential theoretical foundation to understanding the scientific basis of survey research.

The purpose of sampling is to be able to make generalizations about a population based on a scientifically selected subset of that population. Sampling is necessary because it is generally not practical or feasible to seek information from every member of a population. A sample therefore is intended to become a microcosm of a larger universe. However, the question posed in Chapter One asked how a relatively small subset of cases can be used to represent the much larger population from which the subset has been selected. To address and fully understand the implications of this question, it is important to establish the theoretical basis of sampling and its associated assumptions.

Properties of the Normal Distribution

There are certain standard properties of the normal curve that convey how values of the variable are distributed around the mean. The measurement of distance from the mean is calculated in terms of the standard deviation. The *standard deviation* is, as the name implies, a measurement of dispersion around the mean in standardized units. Consequently, no matter what the variable is (for instance, weight, IQ scores, or income), a constant proportion of the total area under the normal curve will lie between the mean and any given distance from the mean as measured in units of standard deviation.

Chapter Five demonstrated the manner in which the standard deviation is calculated for data derived from a sample. The calculation of the true population standard deviation (σ) is similar, but is shown with different symbols for the mean and standard deviation. The formula for the standard deviation of the true population is:

$$\sigma = \sqrt{\frac{\sum (x - \mu)^2}{N}} \tag{6.1}$$

where
    σ = true population standard deviation
    μ = true population mean
    N = population size

These symbols for the mean and standard deviation are known as *population parameters* when they refer to the true population, in contrast to the formulas from Chapter Five that contained symbols for a sample and are known as *sample statistics*.

For any particular normal distribution, regardless of the mean or the calculated standard deviation, the proportion of cases between the mean (μ) and one standard deviation (σ) always turns out to include 34.13 percent of the total cases (see Figure 6.1). Furthermore, since the normal distribution is symmetrical, the identical proportion of cases will lie below the mean (that is, between μ and -1 σ). Hence, 68.26 percent of all cases in the entire population will be found within one standard deviation of the mean in either direction. Similarly, 95.44 percent of all cases are to be found within two standard deviations and 99.74 percent within three standard deviations. It should be clear, therefore, that a very small number of cases exists beyond three standard deviations from the mean.

For example, in a population of twenty-year-old men who have recently completed their military basic training, assume a true population mean weight of 170 pounds and an associated standard deviation of 15 pounds. It can be expected that in that population, 68.26 percent of the males would weigh between 155 and 185 pounds, 95.44 percent would weigh between 140 and 200 pounds, and almost all men (99.74 percent) would weigh between 125 and 215 pounds (see Figure 6.2).

FIGURE 6.1.   AREA UNDER THE NORMAL CURVE.

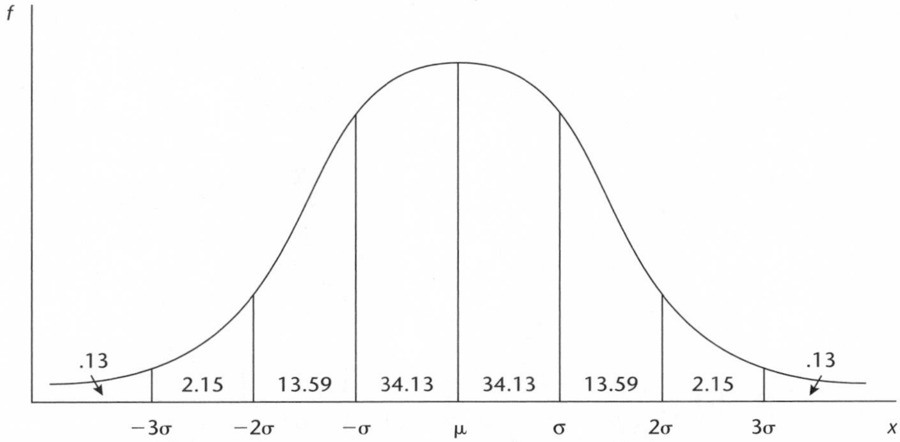

The Standardized *Z* Score

Let it be supposed that one individual in this population of military personnel weighs 176 pounds. It can be readily observed from Figure 6.2 that this individual's weight would lie between the mean and one standard deviation above the mean. Figure 6.2, however, does not provide any greater specificity than that regarding this individual's relative weight. For instance, if this individual wished to know the exact proportion of such military personnel who weigh more than he does, Figure 6.2 would not suffice because his weight is other than 125, 140, 155, 170, 185, 200, or 215 pounds (one, two, or three full standard deviations above or below the mean).

It is possible, however, in the normal distribution, to calculate the relative position of any score by converting it into fractional units of standard deviations, known as standardized Z scores. This conversion can be accomplished through the following formula:

$$Z = \frac{x - \mu}{\sigma} \tag{6.2}$$

where

x = individual score

μ = mean of the population distribution

σ = standard deviation of the population

Z = standard deviation unit scores

FIGURE 6.2.   PROPORTIONATE AREAS UNDER THE NORMAL CURVE FOR MILITARY WEIGHTS.

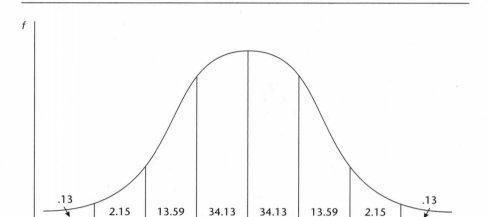

When converting to Z scores, the population mean (in this case, $\mu = 170$) is represented by a Z score of 0 $[(170 - 170)/15 = 0]$. Applying equation 6.2 to the individual's weight of 176 pounds generates a Z score as follows: $(176 - 170)/15 = 0.40$.

For example, a weight of 176 pounds is 0.40 standard deviation units to the right (positive) of the mean ($Z = 0$) on the normal curve. To comprehend this score in the context of relative position in the distribution, it is necessary to find that percentage of cases that are above or below 0.40 standard deviations. That is accomplished by consulting the Table of Areas of a Standard Normal Distribution (given in Resource A). A Z score of 0.40 (column A) represents the point on the curve at which 65.54 percent of all weights are lower than the subject weight of 176 and 34.46 percent are greater. The determination of the percentage of all scores below $Z = 0.40$ is derived from column B in Resource A, which shows 15.54 percent of scores existing between the mean and the Z under consideration. Since the properties of the normal distribution stipulate that 50 percent of all cases are on each side of the mean (see discussion in Chapter Five), adding the 50 percent of cases below the mean to the 15.54 percent of cases above it (column B) yields 65.54 percent.

Another military man who recently advanced from basic training weighs 152 pounds. His relative position in this population is:

$$Z = \frac{152 - 170}{15} = -1.20$$

Negative Z scores reflect the fact that the individual weight under analysis is less than the mean. The determination of the relative position of this Z score also uses Resource A, with an understanding of the symmetrical nature of the normal distribution. A Z score of -1.20 can be evaluated by referring to $Z = 1.20$ and noting that there is a standard percentage of 38.49 percent of cases between the mean and either 1.20 standard deviations above or below the mean. Hence, with the negative Z of -1.20, 88.49 percent (38.49 percent + 50.00 percent) of all individuals in this military population weigh more than 152 pounds, and 11.51 percent weigh less. Figure 6.3 provides a graphic illustration of this example.

If researchers need to identify the proportion of military men in this population who weigh between these individual weights of 152 pounds and 176 pounds, that proportion is the sum of the two percentages from Resource A: 15.54 percent + 38.49 percent = 54.03 percent.

Another valuable application of the standardized Z score is found in the comparative evaluation of the relative position of members of separate, normally distributed populations. Continuing with the military example, assume

FIGURE 6.3.   STANDARDIZED PROPORTIONATE AREAS UNDER THE NORMAL CURVE: AN EXAMPLE.

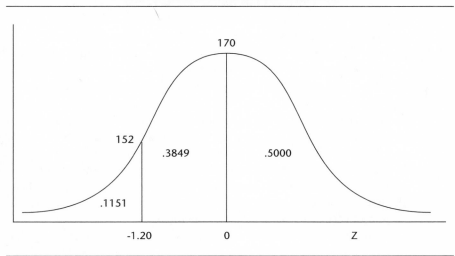

that twenty-year-old women in the military who have recently advanced from basic training have a mean weight of 120 pounds, with a standard deviation of 10 pounds. If a particular woman in this population weighs 127 pounds, it is possible to identify which individual from all of those discussed in this section weighs more relative to his or her own population. The woman's Z score is:

$$Z = \frac{127 - 120}{10} = 0.70$$

This Z score exceeds the Z scores of both men (0.40 and −1.20) and is therefore indicative of the conclusion that although she weighs the least among these three people in absolute terms, the woman military member weighs more relative to her own population than do either of the men.

The Theoretical Basis of Sampling

So far, the discussion has focused on the normal distribution, the true mean, and the true standard deviation of every case in the population. It should be evident that such complete information is rarely available. Gathering data from every member of a population is, in most cases, either logistically impossible or economically infeasible. Therefore, it has become practical for samples of the population to be selected so that generalizations can be inferred from the sample to the total population. These generalizations find their statistical basis in the characteristics of the normal distribution.

As stated in Chapter One, the average person with little statistical background is quite skeptical about the prospect of making generalizations from a single, relatively small sample. Therefore, let it be assumed that in order to determine the mean weight of recently trained twenty-year-old military personnel and to simultaneously appease those who are skeptical, the researcher suggests that 100 separate, mutually exclusive samples be conducted from the same population and that the mean of each of the 100 sample mean weights be calculated in order to estimate the total population's mean weight. The skeptics agree, feeling somewhat more confident of the accuracy of these results compared to those of a single sample. They would be even more comfortable were 1,000 samples used.

The researcher would first select 100 samples (according to principles that will be fully established in Chapters Eight and Nine); he or she would then calculate the mean weight from each of the 100 samples. Table 6.1 presents such sample data, and Figure 6.4 plots these sample means.

The distribution of sample means presented in Figure 6.4 has certain properties that give it a critical role in the sampling process. These properties can

TABLE 6.1. DISTRIBUTION OF 100 HYPOTHETICAL SAMPLE MEAN WEIGHTS.

Sample Means (pounds)	f
175	1
174	1
173	9
172	11
171	16
170	22
169	16
168	12
167	9
166	2
165	1
Total	100

FIGURE 6.4. DISTRIBUTION OF SAMPLE MEANS.

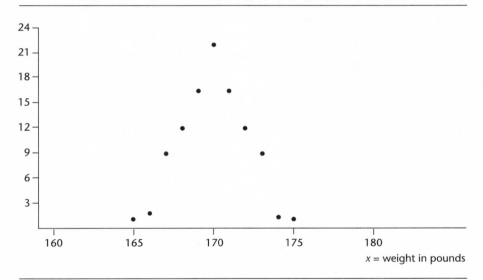

x = weight in pounds

be stated as follows:

Property 1: The value of the mean of sample means ("the mean of means") approaches the true population mean. The larger the number of samples, the closer the approximation is to the population mean. This property is referred to as the *Central Limit Theorem* and is the concept that was agreed to by the skeptical persons above.

Property 2: The distribution of sample means will approximate a normal curve as long as the sample size of each individual sample is sufficiently large (sixty or more at a minimum). This remains true whether or not the raw data are normally distributed.

Property 3: The standard deviation of the distribution of sample means (called the *standard error of the mean* or *standard error*) is smaller than the standard deviation of the total population. There can be a great deal of heterogeneity in the total population. Some males may weigh 100 pounds, others 300 pounds or more. However, when sample means are used, the variation among the mean weights will be significantly less than with the raw data because of the summarizing nature of the mean (Figure 6.5). The standard error is estimated to be:

$$\sigma_{\bar{x}} = \frac{\sigma}{\sqrt{n}} \cdot \sqrt{\frac{\mathcal{N} - n}{\mathcal{N} - 1}} \qquad (6.3)$$

where

$\sigma_{\bar{x}}$ = standard error

σ = population standard deviation

n = size of sampling distributions (number of cases in each sample)

$\mathcal{N}$ = true population size

FIGURE 6.5. HYPOTHETICAL NORMAL DISTRIBUTIONS FOR SAMPLE MEANS COMPARED TO RAW DATA.

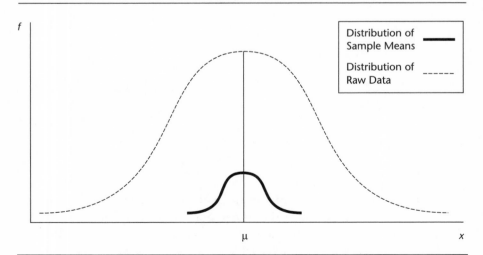

Equation 6.3 takes into account the true population size ($\mathcal{N}$) in the calculation of the standard error. As the population size increases, it can be seen that $\sqrt{(\mathcal{N} - n)/(\mathcal{N} - 1)}$ approaches 1. Hence, with large populations, the standard error approaches

$$\sigma_{\bar{x}} = \frac{\sigma}{\sqrt{n}} \qquad (6.4)$$

The expression $\sqrt{(\mathcal{N} - n)/(\mathcal{N} - 1)}$ has come to be known as the finite population correction. The distinction between the population sizes of large and small populations will be defined in Chapter Seven. At present, assume that the population being analyzed in this chapter is large.

Generalizing from a Single Sample

Sampling theory invokes the three properties discussed above for purposes of justifying the use of one single sample to make inferences about a larger population rather than conducting many separate samples, as the researcher agreed to do above. The process of many samples is quite costly and time-consuming—hence, the desirability of being able to generalize from a single sample.

With regard to the weights of military personnel (Table 6.1), the mean weight ($\bar{x}$) of the sample means is 170. The Central Limit Theorem (Property 1) stipulates that this mean approaches and is a good estimate of the true population mean (μ). Property 2 (assumption of normality) permits us to adapt this information into the context of the normal curve (see Figure 6.6) through the calculation of the standard error (Property 3). The example postulated a population standard deviation of 15 pounds and the size of the samples was 100; hence the standard error of the distribution of the one hundred sample means equals $\sigma_{\bar{x}} = \sigma/\sqrt{n} = 15\sqrt{100} = 1.5$.

In this example, 68.26 percent of all sample means can be expected to fall between 168.5 and 171.5 pounds. In the case of the 100 sample means conducted, therefore, it is expected that 68 or 69 samples would indicate mean weights between 168.5 pounds and 171.5 pounds. Similarly, 95.44 percent of all sample means should lie between 167 pounds and 173 pounds, and 99.74 percent of all sample means should be found between 165.5 and 174.5 pounds.

Another way to look at these sample means is in probabilistic terms. In other words, given a true population mean of 170 pounds and true standard deviation of fifteen pounds, if only one single sample of 100 persons were to be conducted,

FIGURE 6.6. PROPERTIES OF NORMAL CURVES APPLIED TO HYPOTHETICAL DISTRIBUTION OF SAMPLE MEAN WEIGHTS.

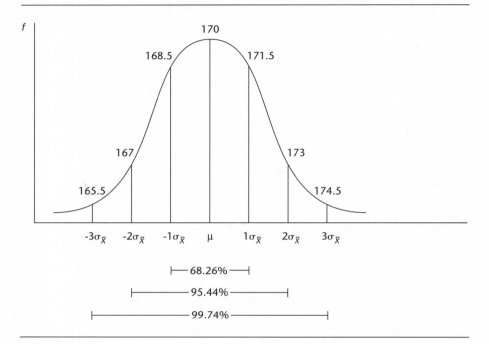

the chances of the sample's mean weight being within the 168.5 to 171.5 pound range is 68.26 percent (.6826 in probability terms), and there is a .9544 probability of the sample mean being within the 167 to 173 pound range. This idea of assessing one single sample in probabilistic estimates of its proximity to the true mean is the essence of sampling.

Sampling affords the researcher the ability to estimate the true mean and standard deviation by understanding the probability of any one sample's likelihood of approximating the true sample mean. It does so through using the normal distribution to understand that any one sample has a 68.26 percent chance of its mean coming within one standard error of the true population mean, a 95.44 percent chance of coming within two standard errors, and a 97.44 percent chance of being within three standard errors. Chapter Seven will apply these probabilities to sample data by determining the range of likely values for the true mean based on the single sample, which ranges are known as confidence intervals.

EXERCISES

1. Suppose that Z is a standard normal variable. Find the proportion of the distribution for Z, which is
   a. above 1.25
   b. below 2.30
   c. below −0.85
   d. between −2.05 and 1.11
   e. between −1.60 and −2.50
   f. between 1.28 and 6.75

2. Suppose that three tests were given in a statistics course that you took. The class averages and the standard deviations were:

Test	Mean	Standard Deviation
1	78	11
2	65	16
3	73	9

   Your scores were 91, 80, and 85 for Tests 1, 2, and 3, respectively. On which test did you do the best (relative to the other students)? On which did you do the worst?

3. The average California household uses 14,000 gallons of water per month, with a standard deviation of 3,000 gallons. Assume a normal distribution.
   a. What proportion of California households uses less than 14,000 gallons of water per month?
   b. What proportion of California households uses more than 8,000 gallons per month?
   c. What proportion of California households uses between 8,000 and 12,000 gallons per month?

4. Sarah received a 470 on the mathematics section of the SAT exam and a 425 on the verbal section. The national mean for mathematics is 510, with a standard deviation of 75. The national mean for the verbal section is 470, with a standard deviation of 90. On which section did Sarah do relatively better in comparison to the national performance?

5. Calculate the standard errors for the following samples:
   a. $\bar{x} = 10$
      $\sigma = 2$
      $n = 50$
   b. $\bar{x} = 3,000$
      $\sigma = 500$
      $n = 700$

6. The record for most home runs hit by a baseball team in one season is held by the Seattle Mariners, who hit 264 home runs in 1997. In 1997, the mean number of home runs hit by each team in baseball was 166, and the standard deviation was 35 home runs. The record had been set the previous year by the Baltimore Orioles, who hit 257 home runs against a league mean and standard deviation of 177 and 40 home runs, respectively.

Prior to Baltimore in 1996, the record had been set by the New York Yankees in 1961 with Roger Maris and Mickey Mantle, who hit 240 home runs when the league average was 151 and the standard deviation was 38. However, maybe the most impressive performance was by the 1927 New York Yankees and their "Murderer's Row," which included Babe Ruth and Lou Gehrig. That team hit 158 home runs when the league average was 58 and the standard deviation was 35.

a. Which of these teams (1927 Yankees, 1961 Yankees, 1996 Orioles, or 1997 Mariners) had the best home run season in relation to the other teams in their respective seasons?

b. In 1997, the San Diego Padres hit 152 home runs. Based on the characteristics of the normal distribution, what proportion of teams did San Diego hit more home runs than? (Use whole number percentages.)

c. Another team fared better than 25 percent of the other teams. How many home runs did that team hit?

CHAPTER SEVEN

CONFIDENCE INTERVALS AND BASIC HYPOTHESIS TESTING

The discussion in Chapter Six made use of population parameters, such as the population mean (μ) and the population standard deviation (σ). However, as discussed, the researcher rarely, if ever, is actually in possession of such information. Probabilistic determinations of the true mean were made using sample sizes and the true population standard deviation, and therein lies a hole in sampling theory. The procedure discussed for estimating μ from one sample requires that σ be known, and if μ is not known, then σ will also be unknown.

Equation 6.4 can be adapted to best accommodate the results of a single sample through the process of substitution. In the absence of σ, the researcher can substitute the standard deviation from a prior comparable study, if available, or, in the absence of a prior comparable study, the standard deviation from the single sample (s) for the population standard deviation (σ). The standard deviation of the prior study or of the sample itself is used as the best available estimate of σ since it is known probabilistically that it is likely that the sample standard deviation will be close to the true population standard deviation and since any differences are mitigated, in large part, by the fact that the standard error reduces the magnitude of any difference between the true standard deviation and the sample standard deviation by Equation 6.4 can be expressed in its single sample form as follows:

$$s_{\bar{x}} = \frac{s}{\sqrt{n}} \tag{7.1}$$

where

    n = sample size (number of cases in a single sample)

    $\bar{x}$ = standard error of the mean for the single sample

    s = sample standard deviation

By mere chance alone, some difference between a sample and the population from which it is drawn must always be expected to exist. In all likelihood, the population mean (μ) will rarely be the same as the sample mean, and the population standard deviation (σ) is highly unlikely to be the same as the sample standard deviation (s). These differences are known as *sampling error,* and they can be expected to result regardless of how scientifically the sample has been selected and implemented. The sampling error that exists between the sample and its population can be formally depicted through the use of *confidence intervals.*

Confidence Intervals

Suppose that in a single sample of 400 university undergraduates, it is found that the mean sample grade point average is 2.75 with a standard deviation (s) of 0.4. Applying equation 7.1, the standard error is obtained as follows:

$$s_{\bar{x}} = 0.4/\sqrt{400} = 0.4/20 = .02$$

It is known that there is .6826 probability that the mean of any sample drawn from a population will be within one standard error of the true mean (Figure 7.1). Hence, the researcher is 68.26 percent confident that μ is in the interval bounded by $\bar{x} \pm s_{\bar{x}}$.

In terms of the specific example:

There is 68.26 percent confidence that $2.75 - .02 \leq \mu \leq 2.75 + .02$. (one standard error).

There is 95.44 percent confidence that $2.75 - .04 \leq \mu \leq 2.75 + .04$. (two standard errors).

There is 99.74 percent confidence that $2.75 - .06 \leq \mu \leq 2.75 + .06$. (three standard errors).

The researcher is 68.26 percent confident that with a sample mean of 2.75 and a sample standard error of 0.02, the true population mean grade point average lies between 2.73 and 2.77. There is 95.44 percent confidence that the true population mean is found within the interval bounded by 2.71 and 2.79 and 99.74 percent confidence that it is between 2.69 and 2.81.

In most scientific investigations, a confidence level of 68.26 percent is not satisfactory. It is common for a researcher to seek a 95 percent or 99 percent level of confidence. The choice of a confidence level is often a trade-off among economy, precision, and risk of error. (The factors associated with the choice of a confidence level will be discussed in Chapters Eight and Nine.) Referring to Figure 7.1 and Resource A, it can be shown that a standardized Z score of ± 2.575 encompasses approximately 99 percent of all cases and that a standardized Z of ± 1.96 includes 95 percent of the cases. Thus, the 95 percent and 99 percent confidence intervals take on the following configurations:

$$(95\%)\,\overline{x} - 1.96 s_{\overline{x}} \leq \mu \leq \overline{x} + 1.96 s_{\overline{x}} \tag{7.2}$$

$$(99\%)\,\overline{x} - 2.575 s_{\overline{x}} \leq \mu \leq \overline{x} + 2.575 s_{\overline{x}} \tag{7.3}$$

The researcher can be 95 percent confident that μ is located in the range expressed by $\overline{x} \pm 1.96 s_{\overline{x}}$ and 99 percent confident that μ is found within $\overline{x} \pm 2.575 s_{\overline{x}}$.

In the above example of undergraduate grade point averages, based on the single sample, we can be 95 percent confident that the true mean undergraduate grade point average is between 2.711 and 2.789 because 95 percent is the probability that the sample mean will be within 1.96 standard errors of the true population mean. Similarly, we can be 99 percent confident that the true mean is between 2.699 and 2.801.

Hence, through the use of confidence intervals, the researcher is able to determine that the true population mean can be estimated within a fixed interval range based on one sample mean. Notice should be taken that for any given sample size, the more rigorous the level of confidence demanded, the more broadly delineated the confidence interval must be. By broadening the confidence interval, the researcher can mitigate the risk of making an error in generalizing from the sample to the population at large.

Confidence Intervals Expressed as Proportions

The reader is more likely to be familiar with the concept of confidence intervals in the context of percentages (proportions). Almost everyone has been exposed to political public opinion polls.

In general, the results, when given in terms of proportions, can be expressed as follows:

$$(95\%)\, p = \bar{p} \pm 1.96\sigma_{\bar{p}} \tag{7.4}$$

$$(99\%)\, p = \bar{p} \pm 2.575\sigma_{\bar{p}} \tag{7.5}$$

where p = true population proportion, $\bar{p}$ = sample proportion, and the standard error of the population proportion is estimated by the sample mean proportion $(s_{\bar{p}})$ as follows:

$$s_{\bar{p}} = \sqrt{\frac{\bar{p}(1-\bar{p})}{n}} \tag{7.6}$$

It is important to note the parallel between equation 7.6 (standard error for proportions) and equation 7.1. The equation for the standard deviation of a proportional distribution is $\sigma_p = \sqrt{p(1-p)}$. Just as with interval data, the equation for standard error requires the standard deviation to be divided by $\sqrt{n}$. Hence, the standard error for a distribution of sample proportions is $\sigma_p = \sqrt{(p(1-p))/n}$. Equation 7.6 represents the standard error for a single sample proportion, where $\bar{p}$ is substituted for p in the same manner as s is substituted for σ in interval data.

The typical results of a public opinion poll might appear as follows:

Twelve hundred scientifically selected respondents were asked to state their preference between Senator Segura and Representative Williams for the office of president of the United States. The results of the survey are as follows:

Segura   47 percent
Williams   45 percent
Undecided   8 percent

The survey contains a margin of error of ± 3 percent.

Such reports rarely contain specific references to confidence intervals and confidence levels, but these results fit precisely into the interpretive context discussed above with regard to interval scale data. The true meaning of this survey finding is that the researcher is 95 percent confident that Segura has between 44 percent and 50 percent of the vote (47 ± 3 percent) and that Williams has between 42 percent and 48 percent. This is comparable to the interval scale example above, when the researcher was 95 percent confident that the mean grade point average was between 2.711 and 2.789.

Let it be assumed that it is much later in the presidential campaign than when this poll was initially taken, and Senator Segura would like to know where he stands now. He commissions a sample survey of 2,000 registered voters, which finds Segura's support to be 52 percent. Encouraged by this finding, but understanding the concept of sampling error, Senator Segura would like to know the margin of error associated with this poll. In other words, can the campaign staff really be confident that Segura possesses majority support? Senator Segura has asked his statistician to respond to that question at a confidence level of 99 percent. Equations 7.5 and 7.6 yield the following result:

$$p = \overline{p} \pm 2.575 s_{\overline{p}}$$
$$= .52 \pm 2.575 \left(\sqrt{\frac{(.52)(.48)}{2,000}} \right)$$
$$= .52 \pm 2.575(.011)$$
$$= .52 \pm .028$$

Thus, Senator Segura can be 99 percent certain that his campaign has between 49.2 and 54.8 percent of the vote. He may be somewhat disappointed that majority support cannot be claimed with 99 percent confidence, but it is better not to be misled by the poll that indicated 52 percent support than it would be to proceed as if majority support were certain. The reader should verify that the 95 percent confidence interval is .52 +/− .022, which might alarm Senator Segura in that it does not seem to indicate that he has even 95 percent confidence of a majority, but this conclusion is premature and will be called into question in the Single and Dual Direction Research Questions section of this chapter.

Hypothesis Testing

The question posed by Senator Segura is more commonly known as *hypothesis testing* based on a sample. The sampling error associated with any sample is used to determine if the sample permits certain conclusions to be drawn from a sample against some standard or hypothetical objective under conditions of 95 percent and 99 percent confidence. That is, a confidence interval identifies the range of the likely location of the true mean and can be used in many cases to answer a question such as Senator Segura's. However, there is a variation on confidence intervals that is better able to address the different types of hypotheses than can confidence intervals. Hypothesis testing is that variation.

Recall the example in Chapter Six in which the mean weight of a 400-person sample of military personnel was reported as 170 pounds, with a standard deviation of 15 pounds. If the Department of Defense had established a mean weight of 165 pounds as indicative of a healthy military population (established standard mean weight), is a researcher 95 percent confident that the population sampled could be characterized as healthy or unhealthy because of its sample mean of 170 pounds?

Rather than determining a confidence interval, a researcher can calculate the number of standard errors that separate the sample mean from the hypothesized or standard mean. That is done by determining the number of standardized Z units between the two means. The formula for Z would be as follows:

$$Z = \frac{\bar{x} - \mu}{s_{\bar{x}}} \tag{7.7}$$

where

$\bar{x}$ = sample mean

μ = established standard mean

$s_{\bar{x}}$ = standard error of the sample mean

$$\frac{s}{\sqrt{n}}$$

where s = sample standard deviation and n = sample size. The denominator $(s/\sqrt{n})$ is the standard error of the sample.

This Z test permits the researcher to assess the health of the military personnel more easily than with confidence intervals. First, the researcher poses a *null hypothesis*, as follows: Null Hypotheis (H_0): There is no difference between the sample mean and the standard, which means that the population is healthy. Second, the researcher applies Equation 7.7:

$$\begin{aligned}
Z &= \frac{\bar{x} - \mu}{s_{\bar{x}}} = \frac{\bar{x} - \mu}{s/\sqrt{n}} \\
&= \frac{170 - 165}{15/\sqrt{400}} \\
&= \frac{5}{0.75} \\
&= 6.67
\end{aligned}$$

Third, the calculated Z of 6.67 must be compared to the two "critical" Z scores that represent 95 percent and 99 percent confidence: 1.96 and 2.575. The comparison is made in absolute value terms in that calculated Z scores with absolute values that equal or exceed these critical values indicate that the null hypothesis can be rejected and that the sample and standard are not the same. This indicates that the sample population is statistically different from a standard healthy population and is, therefore, unhealthy. This difference is also known as *statistical significance*. Statistical significance indicates a sample finding that is sufficiently different from the hypothesized result that sampling error is not able to explain the difference. Under such circumstances, the apparent result can be applied to the population that the sample represents (in this case, military recruits). That is, the sample finding that sampled military recruits do not meet the Department of Defense standard can be "generalized" to all military recruits if the sample has been drawn appropriately (see Chapter Nine). Chapters Ten and Eleven explore additional tests of statistical significance in greater detail.

There is also a form of this test regarding standards and hypothesized true populations that applies to data generated in the form of proportions. As in the case of the interval data, a Z score is calculated and compared to the appropriate critical Z in order to determine if the null hypothesis can be rejected and thereby ascertain a statistically significant finding. The equation for proportions is as follows:

$$Z = \frac{\bar{p} - p}{\sqrt{\dfrac{\bar{p}(1 - \bar{p})}{n}}} \qquad (7.8)$$

where

$\bar{p}$ = sample population mean proportion
p = established standard proportion
n = sample size

To illustrate further, suppose that a sample of 70 faculty members at a midsized liberal arts college in Ohio was polled to find out if the faculty incorporates ethical and moral components in their courses. The State of Ohio's accrediting agency insists that a balanced program of study must include such material in approximately 16 percent of all classes taken. The survey indicated that 21 of the 70 faculty members (30 percent) in the sample included ethical and moral teachings in their classes. Is this finding indicative of a lack of balance in the curriculum (95 percent confidence)? That may appear to be the case, but sampling error must be ruled out. Equation 7.8 can be applied to this situation as follows:

$$Z = \frac{0.30 - 0.16}{\sqrt{\dfrac{0.30(0.7)}{70}}} = \frac{.14}{\sqrt{0.003}} = 2.556$$

Because the calculated Z of 2.556 exceeds the critical Z of 1.96 (95 percent confidence), the researcher can conclude with 95 percent confidence that the school is out of balance with the state's standards. However, were 99 percent confidence required (2.575), that same conclusion could not be drawn.

Single- and Dual-Direction Research Questions

Generally a research question is posed not in terms of an overall difference between the sample and a hypothesized finding (as were the above questions about lack of balance and unhealthy recruits) but rather in terms of a single direction of difference. In the case at hand, instead of posing the research question in the context of general differences between the college and the state standard, the researcher is much more likely to examine the survey results and hypothesize that the college teaches too much ethics and may be shortchanging other disciplines.

A single-direction research hypothesis, such as this, requires a reconfiguration of the confidence interval. When the question is one of difference only, the researcher must allow the possibility of sampling error in both directions. That is, a researcher who is testing to find if there is a significant mean weight difference between military recruits and the standard can reach a conclusion of significant difference if recruits' mean weight is either significantly less than or significantly greater than the standard. Since the researcher can err in either direction, he or she must allocate the possibility of error evenly between these two directions. Figure 7.1 illustrates this allocation of error for the 95 percent confidence level where the

FIGURE 7.1. DUAL-DIRECTION RESEARCH QUESTION (TWO-TAIL TEST AT 95 PERCENT CONFIDENCE LEVEL).

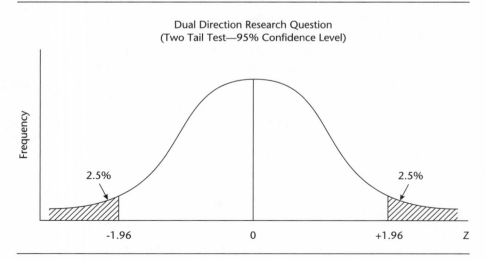

FIGURE 7.2.   SINGLE-DIRECTION RESEARCH QUESTION (ONE-TAIL TEST AT 95 PERCENT CONFIDENCE LEVEL).

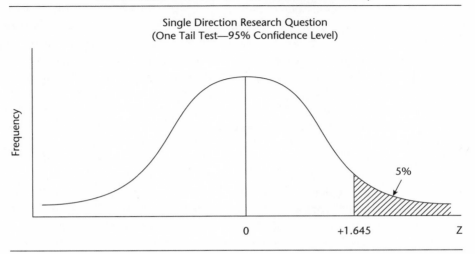

confidence interval is formed at ± 1.96 standard errors from the mean, with the 5 percent error divided equally (2.5 percent in either tail of the diagram).

When the research question is oriented in only one direction, the 95 percent confidence interval shifts totally toward one side of the distribution. Inasmuch as the question is posed in one direction only, the error can be made in one direction only. Therefore, the entire 5 percent is allocated to one tail at 1.645 standard errors from the mean. Figure 7.2 illustrates the confidence interval in the case of a single-direction, 95 percent level of confidence.

Statistical significance for the single-direction hypothesis test therefore can be established at a lower absolute value of Z, but the researcher must be careful to structure the question in a manner that reflects the results of the data. For example, the researcher in the military weight example must pose the question in either of the following formats:

- Are recruits overweight? Look only for a positive Z score.
- Are recruits underweight? Look only for a negative Z score.

With sample data indicating a mean weight of 170 pounds, in contrast to the standard of 165, the appropriate *research hypothesis* is that the recruits are overweight and the null hypothesis is that their weight is not different from the standard— that is, they are not overweight.

Hence for the military recruits, a calculated Z of 6.67 is found to be greater than the single-direction critical Z of 1.645, the null hypothesis can be rejected, and the total population of military recruits can be said to be overweight on average. In the dual-direction situation, critical Z is found at a higher absolute value, and because there is an interest in difference only, the structure of the question need not address which group has a higher or lower measure than the other.

The single-direction critical Z of 1.645 has been derived from Resource A (Table of Areas of a Standard Normal Distribution). You should verify that a Z of 1.645 indicates that 95 percent of the data is on one side of it and 5 percent on the other. For 99 percent confidence levels, the reader should further verify that the single-direction critical Z equals 2.33. Single-direction questions are generally referred to as *one-tail questions*, and dual-direction questions are referred to as *two-tail questions*.

As such, whereas we could not be 99 percent confident that the college in Ohio was out of balance (either positively or negatively) with state ethics standards, the Z score of 2.556 is sufficient to declare with 99 percent confidence that this school spends an overabundance (in contrast to either an overabundance or underabundance) of time in this line of instruction according to state standards.

For Senator Segura's question about a majority, the question really is, "Does Senator Segura have more than 50 percent support?" Using a confidence interval did not account for the one-tail nature of the question. Using hypothesis testing, we can use a one-tail application of equation 7.8, as follows:

$$Z = \frac{.52 - .50}{\sqrt{\dfrac{(.52)(.48)}{2000}}} = \frac{.02}{\sqrt{.000125}} = 1.709$$

Senator Segura is 95 percent confident of a majority ($Z \geq 1.645$), but he has not achieved the 99 percent level of confidence that he desired ($Z \leq 2.33$).

In sum, statistical significance in hypothesis testing can be established when calculated Z equals or exceeds the critical Z scores in Exhibit 7.1.

EXHIBIT 7.1.   VALUES OF CRITICAL Z SCORES FOR RESEARCH HYPOTHESES (ACCORDING TO RESEARCH HYPOTHESIS DIRECTION).

Level of Confidence	Two Directions (Two-Tail)	One Direction (One-Tail)
95 percent	1.96 (absolute value)	1.645 (correct sign only)
99 percent	2.575 (absolute value)	2.33 (correct sign only)

The *t* Test

Recall from Chapter Six that the properties of the normal distribution assume that the sample sizes in the distribution of sample means is sufficiently large—sixty or more. When the sample size is fewer than sixty, as might be the case, for instance, in a random sample of city departmental directors or a random sample of teachers from public schools in one neighborhood, the application of the properties of the normal distribution is problematic. Hence, Z scores are not applicable for such small samples. The *t* distribution, sometimes called Student's *t*, is an adjustment to the normal distribution to account for small sample size, and it should be used in place of Z when the sample size is small (less than sixty).

In hypothesis testing, when the sample size is below sixty, the equation used for determining a calculated Z is applied in almost the same format to produce a calculated *t*, with the only change being the substitution of $n - 1$ for n in the denominator.

$$t = \frac{\bar{p} - p}{\sqrt{\dfrac{\bar{p}(1 - \bar{p})}{n - 1}}} \tag{7.9}$$

The calculated *t* is then compared to the table of critical values of *t* in Exhibit 7.2. The level of confidence, directional properties of the question, and degrees of freedom must be known in order to find critical *t*. Degrees of freedom is a statistical concept that most often represents a derivative of *n*. It can be explained as the number of individual data that are free to vary when only the sum of all data is known. That is, knowing the total number of years of education among parties to a conference of thirty-five people, how many of those individual years of education are free to vary until all others become known? The correct answer is 34 or $n - 1$. In the case of hypothesis testing therefore, degrees of freedom equals $n - 1$. Level of confidence is determined as before in conjunction with the particular research requirements, and the direction of the research question must be determined from the data and from the research objectives.

As with Z, if calculated *t* equals or exceeds the one- or two-tail critical *t* (whichever is applicable), the researcher can conclude that differences between the groups are statistically significant or genuine.

To illustrate the use of the *t* distribution, suppose that 45 percent of a sample of fifty-five teachers at a local high school have indicated that they work in a physically dangerous environment. The school district has established that when 30 percent or more of the teachers in a school feel so threatened, the school is to embark on a costly series of security measures. Can it be concluded that the percentage of the teachers who feel threatened is high enough to require the school to

EXHIBIT 7.2.   CRITICAL VALUES OF THE *t* DISTRIBUTION.

	One-Tail		Two-Tail	
df	$t_{.05}$ (95 percent confidence)	$t_{.01}$ (99 percent confidence)	$t_{.05}$ (95 percent confidence)	$t_{.01}$ (99 percent confidence)
1	6.314	31.821	12.706	63.657
2	2.920	6.965	4.303	9.925
3	2.353	4.541	3.182	5.841
4	2.132	3.747	2.776	4.604
5	2.015	3.365	2.571	4.032
6	1.943	3.143	2.447	3.707
7	1.895	2.998	2.365	3.499
8	1.860	2.896	2.306	3.355
9	1.833	2.821	2.262	3.250
10	1.812	2.764	2.228	3.169
11	1.796	2.718	2.201	3.106
12	1.782	2.681	2.179	3.055
13	1.771	2.650	2.160	3.012
14	1.761	2.624	2.145	2.977
15	1.753	2.602	2.131	2.947
16	1.746	2.583	2.120	2.921
17	1.740	2.567	2.110	2.898
18	1.734	2.552	2.101	2.878
19	1.729	2.539	2.093	2.861
20	1.725	2.528	2.086	2.845
21	1.721	2.518	2.080	2.831
22	1.717	2.508	2.074	2.819
23	1.714	2.500	2.069	2.807
24	1.711	2.492	2.064	2.797
25	1.708	2.485	2.060	2.787
26	1.706	2.479	2.056	2.779
27	1.703	2.473	2.052	2.771
28	1.701	2.467	2.048	2.763
29	1.699	2.462	2.045	2.756
30	1.697	2.457	2.042	2.750
40	1.684	2.423	2.021	2.704
60	1.671	2.390	2.000	2.660
120	1.658	2.358	1.980	2.617
∞	1.645	2.330	1.960	2.575

implement the necessary security? That may appear to be the case, but as always, sampling error must be ruled out by applying the appropriate test of statistical significance. Equation 7.9 can be applied to this situation as follows:

$$t = \frac{0.45 - 0.30}{\sqrt{\dfrac{0.45(0.55)}{54}}} = \frac{0.15}{\sqrt{0.0046}} = \frac{0.15}{0.068} = 2.21$$

Because the calculated t of 2.21 exceeds the one-tail, critical t of 1.676 (95 percent confidence, interpolated between 40 and 60), the researcher can conclude with 95 percent confidence that the school should begin to institute the security measures. Were 99 percent confidence required, that same conclusion could not be drawn (critical $t = 2.402$).

For interval data the t test equation is:

$$t = \frac{\bar{x} - \mu}{\frac{s}{\sqrt{n-1}}} \tag{7.10}$$

EXERCISES

1. The city of Give Me Your Money and Then Leave commissioned a survey of 784 households and found that the mean annual income was \$58,600, with a standard deviation of \$14,000.
   a. Establish the appropriate confidence interval.
   b. What is a confidence interval? Define in terms of its probabilistic characteristics in determining the location of the true population mean.

2. A sample of 1,100 voter registrants was polled to find out whether or not they supported a local initiative placed on the ballot to build a new library. The poll found that 53 percent favored the proposal. Determine the confidence interval at the 99 percent level.

3. A researcher is interested in determining the mean income of attorneys in a particular major metropolitan area. A survey is conducted using a sample of four hundred attorneys, who are selected according to the accepted principles of survey research. It is found that the sample mean income is \$120,000 and the sample standard deviation is \$25,000. The researcher wishes to be 95 percent certain that the mean income reported to the study's sponsor is accurate. Given that the sample mean is subject to sampling error, calculate the appropriate confidence interval.

4. A researcher desires to identify the mean age of full-time four-year college students in Minnesota. A scientific random sample of four hundred respondents indicates a mean age of 26.2 years, with a standard deviation of 5.6 years. If you were asked to interpret this data at a confidence level of 95 percent, what would your confidence interval be?

5. For a city to be eligible for an economic development grant, the federal government requires that the city ensure that the mean household income does not exceed \$25,000. The city is expected to sample its population to estimate its mean income. City X takes a sample of 625 households and finds that the mean household income is \$24,000, with a standard deviation of \$6,500. Can City X claim with 95 percent confidence that it is eligible for a federal grant?

6. Fifteen hundred individuals across the country were asked how many miles they commute to work. The mean response was 5.1 miles, with a standard deviation of 2.0 miles. The Department of Transportation contends that when the average commute exceeds 5.0 miles, the nation is consuming too much oil. Is the nation consuming too much oil according to the Department of Transportation? Be 95 percent confident of your conclusion.

7. A research firm came to a local college and interviewed twenty-six students selected at random. The overall mean SAT score for these students was 1,035, with a standard deviation of 100. Nationally, the mean SAT score is 1,000. Test these data to find out if the local college performs significantly better than the national standard. (Use the 95 percent decision rule.)

8. A sample of five hundred low-income families in a community was surveyed to find out how much these families pay for rent. It was found that the mean monthly per-person rental is $180, with a standard deviation of $30. According to federal law, the community will qualify for federally subsidized housing if the average rental for these families is no more than $165 per person. Using a level of confidence of 95 percent, determine whether the sample can be used as evidence that the community qualifies.

9a. You are interested in evaluating the program of a particular casework agency and have drawn a random sample of 125 cases from its files. The percentage of successfully completed cases is found to be 55. The national standard for such agencies is 60 percent. Can you conclude that the agency is performing differently from that standard (95 percent confidence)?

9b. If the percentage of successful sample cases had been 70, could you have concluded that the agency's performance was better than the national standard? This time, be 99 percent confident of your finding.

10. If Grovers Corners can show that its unemployment rate is 20 percent or greater during the current economic downturn, it will be eligible for federal aid. The application process for this aid is, however, costly and time consuming. Therefore, in order to obtain an indication of Grovers Corners' unemployment situation, a random sample of 400 was selected from those individuals in the workforce. It was found from this sample that the local unemployment rate is 16 percent. Assuming a decision rule of 95 percent, decide whether Grovers Corners should apply for the aid or accept this sample as evidence that the community will not qualify (that is, is the unemployment rate too low?).

11. A sample 200 graduate students at SDSU was surveyed in order to find out how many hours per week they study. It was found that the mean number of hours per week per class is 3.5 (s = 1.1). The National Association of Study Time Issues (NASTI) has declared that the national goal is 3.7 hours per class. Are SDSU professors too easy on their students (99 percent confidence)?

12. Joe Munderotz, Planning Director of the City of Munderotz in Munderotz County, would like to know if there is popular support for his growth

management strategy, which has received widespread publicity in the *Munderotz Messenger* and other local media (known locally as "Mundermedia"). In consequence, he has decided to survey 800 resident adults at random throughout the city to identify the extent of such support. It is found that 51 percent support the concept of growth management, but the Planning Commission, headed by his mother, demands that he have 57 percent support in order to implement the plan. Should Munderotz accept the survey results as evidence that the plan does not meet Planning Commission standards? Use a decision rule of 95 percent.

13. The faculty in the College of Nonprofessional Activity at a large state university in the Western United States was surveyed in order to determine whether they receive "psychic rewards" from teaching. The proportion of faculty responding "yes" was 70 percent with a sample size of twenty-six. The official state poll (used as a standard in these matters) registered a proportion for all state faculty of 52 percent responding "yes". Using a 95 percent confidence level, determine whether the faculty is consistent with the state standard response rate.

14. The Psychological School for the Preparation of Sensitive Public Servants (PSPSPS) conducted a survey of its 2005 graduates to find out how many interviews they went on before securing their current job. A sample of 18 was polled and a mean of 8 interviews was indicated with a standard deviation of 4. The school has long prided itself on its ability to save its students from the headache of such interviews and promises in its brochures that the average number of job interviews per graduate before finding an "exciting and fulfilling position" is 6. Should the school accept this survey as a statistically significant indicator that their brochure might be misleadingly overstated (95 percent decision rule)?

CHAPTER EIGHT

DETERMINING THE SAMPLE SIZE

A crucial question at the outset of a survey research project is how many observations are needed in a sample so that the generalizations discussed in Chapter Seven can be made about the entire population. The answer to this question is by no means clear-cut; it requires the careful consideration of several major factors. Generally, the greater the level of accuracy desired and the more certain the researcher would like to be about the inferences to be made from the sample to the entire population, the larger the sample size must be.

Determinants of Sampling Accuracy

There are two interrelated factors that the researcher must address with specificity before proceeding with the selection of a sample size: *level of confidence* and *confidence interval*. The level of confidence is the risk of error the researcher is willing to accept in the study. Given time requirements, budget, and the magnitude of the consequences of drawing incorrect conclusions from the sample, the researcher will typically choose either a 95 percent level of confidence (5 percent chance of error) or a 99 percent level of confidence (1 percent chance of error). As discussed in Chapter Seven, the confidence interval determines the level of sampling accuracy that the researcher obtains. In this chapter, it will be shown that selection of the sample size is a primary contributor to the researcher's success in achieving a certain degree of sampling accuracy. In other words, sample size is directly related to the accuracy of the sample mean as an estimate of the true population mean.

Recall that the equations for the standard error for samples containing interval scale variables (equation 7.1) or proportions (equation 7.6) included a factor (n) representing sample size. For any given sample standard deviation, the larger the sample size is, the smaller is the standard error. Conversely, the smaller the sample size is, the larger is the standard error. For instance, if a sample of 100 respondents indicates a mean income of $20,000 per year with a sample standard deviation of $3,000, the standard error and associated confidence intervals, with 95 percent or 99 percent levels of confidence, would be calculated as:

$$s_{\bar{x}} = \frac{s}{\sqrt{n}} = \frac{3,000}{\sqrt{100}} = \frac{3,000}{10} = \$300$$

95 Percent Confidence Interval   *99 Percent Confidence Interval*

$\bar{x} \pm 1.96 s_{\bar{x}}$ $\bar{x} \pm 2.575 s_{\bar{x}}$

$20,000 \pm 1.96(\$300)$ $20,000 \pm 2.575(\$300)$

$20,000 \pm \$588$ $20,000 \pm \$773$

Hence, the researcher can be 95 percent certain that the true mean income for the population is between $19,412 and $20,588 or 99 percent certain that the true mean is between $19,227 and $20,773. To phrase it another way, the researcher will have achieved a *margin of error* of ±$588 with 95 percent confidence and ±$773 with 99 percent confidence.

If the available sample contains 400 responses, the standard error and associated confidence intervals would be calculated as

$$s_{\bar{x}} = \frac{\$3,000}{\sqrt{400}} = \$150$$

95 Percent Confidence Interval   *99 Percent Confidence Interval*

$20,000 \pm 1.96(\$150)$ $20,000 \pm 2.575(\$150)$

$20,000 \pm \$294$ $20,000 \pm \$386$

Therefore, with a sample size of 400 rather than 100, the researcher has been able to narrow the margin of error by 50 percent for each level of confidence, respectively, but this 50 percent narrowing has required that the sample size be quadrupled. If the sample size were to be increased to 1,000, the 250 percent increase in sample size from 400 would reduce the margin of error by only an additional 37 percent. It is noteworthy that such reductions in confidence intervals can in fact be achieved, but at the potentially high cost of a substantially larger sample size.

In terms of proportions, a sample of approximately 1,050 persons provides 95 percent confidence of a margin of error of ±3 percent, but 1,850 respondents for 99 percent confidence (more than a 75 percent increase in sample size).

The process of selecting a sample size requires that the researcher determine an acceptable range of uncertainty or margin of error, given the time and cost constraints of the study. In the former example, for instance, a researcher who determines that the study can tolerate no more than a $300 margin of error in either direction from the sample mean would not be satisfied by the 100-person sample at either level of confidence but would be served by the 400-person survey with 95 percent confidence only. In the case of the latter example, the researcher must decide whether increasing the cost and time involved in the study would be justified by an increase in confidence from 95 percent to 99 percent.

There are no fixed criteria by which to make this choice. The researcher must make this determination on a case-by-case basis and in accordance with the particular goals and objectives of the study. This interrelationship among level of confidence, margin of error, and the effect of sample size on these parameters makes the determination of sample size an absolutely vital component of the sample survey process.

The researcher should consider the following guidelines in the selection of sample size:

- The researcher will generally find a margin of error of ±3 to ±5 percent to be satisfactory for proportional data. Interval data margins of error must be established on a case-by-case basis, depending on the unit of measurement, magnitude, and range of the particular variable.
- The greater the consequences are of generating data that might lead to incorrect conclusions, the greater the level of confidence the researcher should establish. In practical terms, this involves a choice between the 95 percent and 99 percent levels of confidence.
- In most cases, the researcher can be satisfied by choosing the 95 percent confidence level, which implies a 5 percent risk that the confidence interval is incorrect.

This choice between 95 percent and 99 percent confidence levels involves a variety of factors, including cost, time, and difficulty of achieving required sample sizes. Assuming, however, that no such constraints exist, one might ask why 99 percent should not be the optimal confidence level choice. To address this question, the concepts of Type I and Type II errors must be discussed.

A *Type I error* is the error associated with making a decision based on the data from the sample. It is the error that this book has discussed as the confidence level. The other form of error associated with sample data is known as a *Type II error;* it derives from being overly conservative and not acting on the results, only to find out later that the researcher should have acted.

To illustrate, suppose that a researcher wanted to administer a new serum to 2,000 randomly selected individuals suffering from a communicable disease. Also suppose that this researcher requires a 99 percent confidence level to conclude that the serum is beneficial and to recommend its mass production and distribution (Type I error = 1 percent). The results of the research showed that there was only 95 percent confidence that patients benefited from the serum; consequently, the researcher could not conclude with 99 percent confidence that the serum should be marketed, and its development was not pursued.

Over the ensuing years, thousands of individuals continued to die of the disease. Eventually another researcher decides to repeat the test on another sample population, this time using a decision rule of 95 percent confidence (Type I error = 5 percent). The results again showed a 95 confidence, but under the new confidence level, this researcher could advocate production of the serum.

Deaths from the disease dropped dramatically over the next few years, demonstrating that a Type II error (not acting on the data after the first experiment) had been made by being overly conservative in setting the original confidence level at 99 percent.

It is now fairly well accepted in the research community that in most instances, the 95 percent level of confidence represents a reasonable balance between the risks associated with Type I and Type II errors.

Determination of Sample Size for Variables Expressed in Terms of Proportions

Determination of sample size for data given in terms of proportions is somewhat more straightforward than when the variable is on an interval scale. Hence, we first introduce this methodology.

Large Populations

The relationship among the confidence interval, the level of confidence, and the standard error of sample proportions can be expressed by the following equation:

$$ME_p = \pm Z_a(\sigma_p) \tag{8.1}$$

where

ME_p = margin of error in terms of proportions
Z_a = Z score for various levels of confidence (α)
σ_p = standard error for a distribution of sample proportions

The formula for the standard error of the true population proportion is $\sigma_p = \sqrt{(p(1-p))/n}$; substituting it into equation 8.1, we can rewrite the equation as follows:

$$ME_p = \pm Z_a \sqrt{\frac{p(1-p)}{n}} \tag{8.2}$$

Solving for n yields

$$n = \left(\frac{Z_a \sqrt{p(1-p)}}{ME_p} \right)^2 \tag{8.3}$$

To proceed with the calculation of specific sample sizes (n), the values of Z_a, ME_p, and p must be established. As discussed, Z_a is most commonly set at 1.96 for the 95 percent level of confidence or 2.575 for 99 percent. The margin of error ME_p is typically set not to exceed 10 percent and is much more frequently set in the 3 to 5 percent range, depending on the specific degree of accuracy to which the findings must conform. As we have seen, the true proportion (p) is unknown, so we have been using the sample proportion, which is unknown prior to actually conducting the survey. The most conservative way of handling this uncertainty for purposes of setting a sample size is to set the value of p at the proportion that would result in the highest sample size. This occurs when $p = .5$; equation 8.3 can be further refined, therefore, to read:

$$n = \left(\frac{Z_a(.5)}{ME_p} \right)^2 \tag{8.4}$$

because $\sqrt{.5(1-.5)} = .5$.

Now suppose that a government decision maker is in the process of determining an appropriate sample size for a study of public opinion concerning community service system adequacy. The question to be posed is whether the respondents find community services to be adequate. Percentages responding yes and no are to be tallied and presented for review. For purposes of this study, the decision maker feels that it is important for the sample proportion to be accurate within ±4 percent of the true proportion, and it is felt that 95 percent confidence in these findings would be satisfactory in order for the information to be effectively used. To obtain the appropriate sample size for this study, the researcher can substitute numbers into equation 8.4 as follows:

$$n = \left[\frac{(1.96)(.5)}{.04} \right]^2 = 600.25$$

The calculated n must be rounded to the next highest whole number, so in this case, a sample size of 601 persons is required.

Keep in mind that $Z_a = 2.575$ can be substituted for $Z_a = 1.96$ when 99 percent confidence is required and that the confidence interval (ME_p) also can be varied according to the researcher's requirement for various levels of sampling accuracy.

For each level of confidence (95 percent or 99 percent), required sample sizes can be calculated for various confidence intervals in terms of proportions by operationalizing equation 8.4. Table 8.1 portrays the calculated required sample sizes under these conditions.

TABLE 8.1. MINIMUM SAMPLE SIZES FOR VARIABLES EXPRESSED AS PROPORTIONS.

	Sample Size	
Margin of Error	95% Confidence	99% Confidence
±1	9,604	16,577
±2	2,401	4,145
±3	1,068	1,842
±4	601	1,037
±5	385	664
±6	267	461
±7	196	339
±8	151	260
±9	119	205
±10	97	166

It should be noted, once again, that there is an important inverse relationship between the sample size and the standard error, as manifested in the margin of error. To narrow the margin of error, an increase in the sample size is required—an increase that can quickly become prohibitively expensive.

Small Populations

As discussed in Chapter Six, sampling theory and the equations derived from it assume a large population size. Therefore, the assumption underlying the calculation of sample sizes in Table 8.1 is that the general population from which the sample or samples are taken is large. If, however, the population is not large, the standard error must be recomputed with the finite population correction included. The formula for sample size in this case becomes:

$$n = \left(\frac{Z_a \sqrt{p(1-p)}}{ME_p} \cdot \sqrt{\frac{N-n}{N-1}} \right)^2 \tag{8.5}$$

Having introduced a factor for n on each side of the equation, we must solve for n again. Doing so yields:

$$n = \frac{Z_a^2 [p(1-p)]N}{Z_a^2 [p(1-p)] + (N-1)ME_p^2} \tag{8.6}$$

Replacing p with .5, as discussed previously, the general equation for sample size in all populations—both large and small—becomes

$$\frac{Z_a^2 (.25)(N)}{Z_a^2 (.25) + (N-1)ME_p^2} \tag{8.7}$$

In practice, since the finite population correction approaches 1 in large populations, this adjusted sample size determinant is used only when populations are not large, and we continue to use equation 8.4 and Table 8.1 when the population is large.

The distinction between "large" and "not large" will be addressed shortly, but prior to doing so, an example of the required sample size from a small population is in order. If a researcher seeks to determine the political party preferences of the 2,500 professors at a large state university and does not have the time or financial resources to interview them all, a sample of professors can be taken. The researcher must decide how many professors to survey and establishes that a 95 percent level of confidence will be satisfactory along with a margin of error that does not exceed ± 3 percent. Applying equation 8.7, the following is obtained:

$$n = \frac{(1.96)^2(.25)(2,500)}{(1.96)^2(.25) + 2,499(.03)^2}$$

$$= \frac{(3.8416)(.25)(2,500)}{(3.8416)(.25) + 2,499(.0009)}$$

$$= \frac{2401}{.9604 + 2.249}$$

$$= 749$$

If the researcher wished to be 99 percent confident of the ±3 percent margin of error, the following sample size would be required:

$$n = \frac{(2.575)^2(.25)(2,500)}{(2.575)^2(.25) + 2,499(.03)^2}$$

$$= 1,061$$

Comparing these results to those in Table 8.1, it can be seen that in the case of this relatively small population, the researcher can obtain 95 percent confidence interviewing 749 professors instead of 1,068 in a large population, as indicated in Table 8.1, and 99 percent confidence can be achieved with 1,061 interviews instead of 1,844. These differences arise from the logical fact that fewer interviews are needed from a very small population in order for that population to be adequately represented by the sample.

Table 8.2 reflects the application of equation 8.7 for various population sizes for the 95 percent and 99 percent levels of confidence and for margins of error of ±3 percent, ±5 percent, and ±10 percent. Notice that as the population size (N) reaches 100,000, the required sample size approaches those listed in Table 8.1. Hence, a population size of 100,000 or greater can be considered large, and populations of fewer than 100,000 could be considered small. For particularly conservative confidence intervals and very small population sizes (designated by note a in Table 8.2), the assumption of normality does not apply. Very small populations are most accurately characterized in terms of a hypergeometric distribution, which is an advanced concept not addressed in this book. In certain of these hypergeometric cases, equation 8.7, which is derived from properties of the normal distribution, does not yield appropriate sample sizes. In lieu of this equation, a sample size of 50 percent of the population size has been determined to provide the required accuracy as long as the sample appears to be relatively representative of the general population

TABLE 8.2.   MINIMUM SAMPLE SIZES FOR SELECTED SMALL POPULATIONS.

	Sample Sizes					
	95% Level of Confidence			99% Level of Confidence		
Population Size (N)	±3%	±5%	±10%	±3%	±5%	±10%
500	250[a]	218	81	250[a]	250[a]	125
1,000	500[a]	278	88	500[a]	399	143
1,500	624	306	91	750[a]	460	150
2,000	696	323	92	959	498	154
3,000	788	341	94	1,142	544	158
5,000	880	357	95	1,347	586	161
10,000	965	370	96	1,556	622	164
20,000	1,014	377	96	1,687	642	165
50,000	1,045	382	96	1,777	655	166
100,000	1,058	383	96	1,809	659	166

Note: The choice of ±3 percent, ±5 percent, and ±10 percent for confidence intervals is based on the tendency of researchers to use these intervals or a similar range of intervals in the design of their surveys.

[a]Population sizes for which the assumption of normality does not apply; in such cases, the appropriate sample size is 50 percent of the population size.

(Yamane, 1967). In sum, the survey administrator will never require a sample size in excess of 50 percent of the total population that the sample represents. This rule has particular significance when it is necessary to undertake internal surveys of organizations with a small population base. Examples of such surveys are employee satisfaction polls and job performance evaluations. It must only be established that these very small samples do an adequate job of representing the full population base.

Determination of Sample Size for Interval Scale Variables

If some sample data are in the form of interval scale variables, some adaptations to the equation are necessary.

Large Populations

For large populations, equation 8.1 must be adapted as follows:

$$ME_i = \pm Z_a(\sigma_{\bar{x}}) \tag{8.8}$$

where

ME_i = confidence interval in terms of interval scale

Z_a = Z score for various levels of confidence (α)

$\sigma_{\bar{x}}$ = standard error for a distribution of sample means

Using equation 6.4 ($\sigma_{\bar{x}} = \sigma/\sqrt{n}$) and substituting it into equation 8.8 results in the following equation:

$$ME_i = Z_a\left(\frac{\sigma}{\sqrt{n}}\right) \tag{8.9}$$

Solving for n yields

$$n = \frac{Z_a^2\sigma^2}{ME_i^2} \tag{8.10}$$

To proceed with the calculation of specific sample sizes (n), the values of Z_a, ME_i, and σ must be established. As discussed, Z_a is most commonly set at 1.96 or 2.575; ME_i is generally set in the context of the variable under study, as the research study dictates, and σ is estimated from the sample data themselves by s, the standard deviation of the single sample, as discussed in Chapter Six. Hence:

$$n = \frac{Z_a^2 s^2}{ME_i^2} \tag{8.11}$$

Suppose that a researcher is interested in obtaining a sample from a large population of households in County X to determine the mean household income. The goals are to select a sample size that will yield a margin of error of no more than ±\$1,000 and to be 95 percent confident of this result.

A problem exists, however, that is similar to the problem that was addressed with regard to proportions by setting $p = .5$, because the value for the sample standard deviation is unknown before the sample is conducted. Furthermore, there exists an additional difficulty for interval data from which proportional data is exempt. That is, there is no simple parameter to use in advance for s that would maximize the determination of n, as there is with proportions (namely, $p = .5$). Since it is not at all likely that accurate information about the population parameters will be known before the completion of the survey, only a reasonable estimate of s can be made. There are two alternative methods of making this estimate:

1. The researcher may wish to use any previous information that is available (for example, another survey of this population that can provide the necessary mean and standard deviation for a key variable).
2. A pilot survey or pretest (see Chapter Two), conducted on the population, will yield a standard deviation that can be used as estimates for the proposed sample.

Because a pretest should be conducted as a critical part of the survey process in any case, it is generally more feasible and efficient to use this method for the estimate of the sample standard deviation.

Based on a pretest of households in County X, a preliminary estimated mean of \$30,000 and standard deviation of \$6,000 are determined. The researcher may now use this information to operationalize equation 8.11 to yield the following sample size:[1]

$$n = \frac{(1.96)^2 (6,000)^2}{(1,000)^2} = \frac{138,297,600}{1,000,000} = 139$$

Small Populations

As is the case when calculating sample sizes for variables in the form of proportions, an adjustment must be made to equation 8.11 when the general population is small (under 100,000). The finite population correction regarding proportions, $\sqrt{(N - n)/(N - 1)}$ is also applicable to interval scale variables. Equation 8.11 becomes

$$n = \left(\frac{Z_a^2 s^2}{ME_i^2} \right) \left(\sqrt{\frac{N - n}{N - 1}} \right) \tag{8.12}$$

which can be written as

$$n = \frac{Z_a^2 s^2}{ME_i^2 + \dfrac{Z_a^2 s^2}{N - 1}} \tag{8.13}$$

In the example regarding household income in County X, suppose that the administrator wishes to research household income in only one small community within the county. This community has a population of 5,000 people. The researcher, as before, is interested in determining a sample size that will yield a margin of error of no more than $1,000 and wishes to be 95 percent confident of the result. A pretest has estimated a mean of $30,000 and a sample standard deviation of $6,000. The researcher may now use this information to operationalize equation 8.13 to yield the following sample size:

$$
\begin{aligned}
n &= \frac{(1.96)^2(6,000)^2}{(1,000)^2 + \dfrac{(1.96)^2(6,000)^2}{4,999}} \\[2ex]
&= \frac{(3.8416)(36,000,000)}{1,000,000 + \dfrac{(3.8416)(36,000,000)}{4,999}} \\[2ex]
&= \frac{138,300,000}{1,000,000 + 27,665.05} \\[2ex]
&= 135
\end{aligned}
$$

Note that as with proportions from small populations, fewer respondents are required from the smaller community than from the entirety of County X to attain the same level of confidence and margin of error. Assuming the requisite characteristics of the normal distribution, this difference in sample size should be intuitively clear. Because interval scale variables tend to provide an enormous and unbounded range of possible values, tables of minimum sample size, such as Tables 8.1 and 8.2 for proportions, are not feasible for the interval scale. Equations 8.11 and 8.13 must be applied in each case. Figure 8.1 summarizes these four distinct determinations of sample size.

FIGURE 8.1.   DETERMINING SAMPLE SIZE.

| | | Margin of Error | |
		Proportion	Interval
Population Size	Large	Equation 8.4 or Table 8.1	Equation 8.11
	Small	Equation 8.7 or Table 8.2	Equation 8.13

Determination of Sample Size When Both Proportional and Interval Scale Variables Are Present

In most surveys, both proportional scale variables and interval scale variables are present. For example, the researcher is usually interested in determining the proportion of respondents who are identified as male or female and may also be interested in knowing what proportion of respondents intend to vote for a particular candidate. At the same time, he or she often wishes to obtain knowledge regarding respondents' household income, age, or years of formal education. These latter variables are expressed in the form of the interval scale. It may be difficult to ascertain which equation or equations to use in order to determine the appropriate sample size for such a survey. The researcher must make certain that the largest sample size requirement is satisfied. Hence, for all interval scale variables and proportional variables with varying margins of error or levels of confidence, the researcher must calculate the required sample sizes by using the appropriate equations or tables and establish a survey sample size equal to the largest required individual variable sample size.

For example, if the researcher calculates that the age variable requires 300 respondents to satisfy the research needs, the income variable requires 350 respondents, and the proportional variables all require a total sample size of 385 respondents, the researcher would conclude that at least 385 respondents must be secured and that this represents the overall survey sample size.

In most instances, to avoid this repetitive and laborious process, a survey administrator will use the proportional variables to select a sample size with an overall margin of error and level of confidence. This sample size generally satisfies the most stringent requirement of the interval scale variables. Be aware, however, that if a particular interval scale variable is crucial to the focus of the survey, it would be prudent to make certain that the general sample size also meets the requirements for that particular variable.

EXERCISES

1. A statewide public opinion survey of 200 respondents is being conducted by Candidate Jones. Jones wishes to be 95 percent confident that the margin of error is ±5 percent. Is the sample size sufficient to satisfy Jones's requirements? If not, what minimum sample size does Jones need?
2. Calculate the minimum required sample size for the following confidence levels and margins of error for a survey conducted in a city with a population of forty thousand.

	Margin of Error (%)	Confidence Level (%)
a.	±4	95
b.	±6	99
c.	±9	95
d.	±2	99

3. Calculate the minimum required sample size for the following confidence levels and margins of error for data gathered from surveys conducted in New York State.

		Margin of Error	Confidence Level (%)
a.	Mean height of adult males ($s = 3$ inches)	±0.5 inch	95
b.	Mean age of entire population ($s = 15$ years)	±2 years	99
c.	Mean income of teachers ($s = \$11,000$)	±$1,000	95

4. Determine sample sizes for the following survey situations:
   a. Resident population of 1 million; confidence level of 99 percent; confidence interval of ±2 percent
   b. Resident population of 45,000; confidence level of 95 percent; confidence interval of ±5 percent
   c. Resident population of 95,000; confidence level of 99 percent; standard deviation of 14 miles; margin of error of ±2 miles

5. City public health officials are seeking to identify the need for a better immunization program. They want to conduct a sample in their city (population = 250,000) that will enable them to identify the number of underimmunized residents within a margin of error of ±57 persons (95 percent confidence level).
   a. How would you proceed to establish the necessary standard deviation prior to conducting the actual survey?
   b. How many city residents must be sampled in order to achieve this desired degree of accuracy given that the standard deviation is 285 persons and the mean is 8,348 persons?

Note

1. The interval scale sample size of 139 respondents may strike some people as somewhat small in comparison with the sample sizes discussed in this chapter. A sample of 139 respondents would correspond to a margin of error of around ±8 percent. The $1,000 margin of error in this example would seem to be less than 8 percent, based on the $30,000 mean. Be aware, however, that the relationship of importance is not between the margin of error and the mean but rather between the margin of error and the standard deviation.

In terms of proportions, the standard deviation is assumed to equal 0.5. The ratio of the margin of error to the standard deviation in this example is $0.08/0.5 = 0.16$. The interval margin of error can be assessed similarly, generating the same ratio of margin of error to standard deviation:

$$\frac{\text{margin of error}}{\text{standard deviation}} = \frac{1{,}000}{6{,}000} = 0.167$$

CHAPTER NINE

SELECTING A REPRESENTATIVE SAMPLE

Although sample size is very important, it is by no means the only determinant of what constitutes adequacy of the sample. It is critical that the sample be selected according to well-established, specific principles. The purpose of this chapter is to discuss these principles and the methods of sample selection that have been adapted and derived from them.

Probability Sampling

Sampling methods can be categorized into probability sampling and nonprobability sampling. In *probability sampling*, the probability that any member of the working population will be selected to be part of the eventual sample is known. This implies extensive and thorough knowledge of the composition and size of the population. In *nonprobability sampling*, the selection process is not formal; knowledge of the population is limited, and hence the probability of selecting any given unit of the population cannot be determined.

There are two characteristics of probability samples:

- The probability of selection is equal for all members of the population at all stages of the selection process.
- Sampling is conducted with elements of the sample selected independently of one another (one at a time).

For example, consider a population that consists of 100 persons whose names are written on equal-sized pieces of paper and placed in a hat for selection. The pieces of paper are thoroughly mixed and selected one by one, without being seen, until the sample size is obtained. This is a probability sample. Of course, when a piece of paper is selected from the hat to become part of the sample, the remaining papers no longer have the same probability of being selected as the previous one did. For instance, each member of a population of 100 has a 1–in–100 chance of being selected. After 25 have been selected, however, those remaining have a 1 in 75 chance of being selected. This might seem to violate the first rule of probability sampling (equal probabilities), but as long as the chances for selection are equal at any given stage in the sampling process, the rule is not violated. This is called *sampling without replacement,* and it is particularly acceptable when the population is relatively large, because the probability differences from stage to stage are negligible.

Identification of the Sampling Frame

The first consideration in deriving a sample is specification of the unit of analysis. The *unit of analysis* is the individual, object, or institution (or group of individuals, objects, or institutions) that bears relevance to the researcher's study. This relevance relates to the concept of a general universe or population. The *general population* can be defined as the universe to which the research findings will be applied. In other words, it is the theoretical population to which the researcher wishes to generalize the study's findings. This population is composed of units, which become the units of analysis for purposes of the study. Specifically, a unit of analysis can be a person, a household, a social organization, a political jurisdiction, a corporation, an industry, a hospital, or a geographical entity, among others, depending on the nature of the conceptual general population. If the objective of the research, for example, is to generalize findings to residents of California, then the units of analysis are individuals. If the study seeks to determine information on a county-wide basis about crime, then counties are the units of analysis.

The designation of a general population, although critical for conceptualizing the purpose and objectives of a study, is generally not conducive to the actual selection of a sample. Let us suppose that a study intends to generalize its findings to the residents of an entire metropolitan area. The general population can be considered to be those people who reside within the political boundaries of the area under study, and the units of analysis are the individual residents.

The process of selecting a representative sample requires, at its theoretical optimum, that the researcher know where and how to contact each person in the population. From a practical standpoint, it is highly unlikely that all members of the general population can be identified and contacted. People die and are born every minute; some people live in unrecorded accommodations (for example, the homeless), and others cannot be contacted for various personal reasons. Herein lies the essence of the statistical necessity for establishing an intermediate step between the general population and the actual sample: the development of a *working population*. The working population is an operational definition of the general population that is representative of the general population and from which the researcher is reasonably able to identify as complete a list as possible of members of the general population. This derivation of a sample from the general population through the use of a working population is known as the *sampling frame*.

Establishing a Representative Working Population

It is useful to illustrate the relationship between the general population and the working population. Suppose that a researcher is studying factors associated with economic disadvantage and poverty in New York City. The general population is to consist of all the economically disadvantaged people in New York City. There are no lists of such a population; therefore, the researcher must substitute a working population for this general population; he or she must find some other identifiable population that can be claimed to correspond to the general population closely enough to be considered its surrogate. The designation of a working population requires the researcher to operationalize the concept of economic disadvantage. There are a variety of definitions of economic disadvantage. The federal government has established poverty-level criteria; poor families receive government aid from various federal, state, and local programs; and other agencies have their own measure of economic disadvantage. Based on any one or any combination of these definitions, a list of persons that constitutes the sampling frame can conceivably be obtained.

None of these lists, either alone or in combination, will be complete and exhaustive in nature. Recognizing this, the researcher will want to select a list, or combination of lists, that maximizes representativeness and minimizes systematic omissions (frequently referred to as *systematic biases*). *Systematic omissions* are absences from the working population of groups that clearly differ from those in the working population. For example, in political polling, people who do not have land-based telephones (for instance, those with only cellular phones), people who screen their incoming phone calls, and people without phones are generally not included in the poll. To the extent that these groups possess significant differences from the people who are contacted, there is a systematic omission in the sample that must be corrected. If the individuals who are omitted are very similar to those who are included in all or most of the important categories (for example, in ethnicity, age, and sex), their omission would not be considered to be systematic and would require no corrective action.

It is usually impossible to eliminate all omissions from the working population. Ideally, the working population would be a complete list of members of the general population. However, no list can be expected to be perfectly complete, and it is therefore important for the researcher to be reasonably satisfied that the working population and derivative sample represent the general population as closely as possible.

In constructing the sampling frame, therefore, the researcher should attempt to determine the extent to which members of the working population have been excluded from this list and decide whether these excluded members differ in any significant manner from those who are included on the list.

In the economically disadvantaged example, the researcher may legitimately choose to define the economically disadvantaged as consisting of households with annual incomes below the federal poverty standard. This definition identifies the working population. The researcher might then identify a working population consisting of individuals who receive funds through the Temporary Aid to Needy Families program (TANF), which requires, for qualification, a household income below the federal poverty level. The researcher is likely to encounter a systematic omission in this list, however, that is sufficient to render it unrepresentative: individuals who may be economically disadvantaged but do not have children are not eligible for the TANF program. However, in this case, since TANF serves only families with children, the list does not include individuals or families without children whose income is sufficiently low to otherwise qualify. Such an omission is clearly systematic in the sense that there is a clear and important difference between those who are included and those who are excluded. A systematic omission of this nature renders the working population unrepresentative of economically disadvantaged persons and therefore unacceptable. The researcher must search for an alternative working population—another list (or combination of lists) of people with incomes below the poverty level that does not contain such a systematic omission or weight the responses (as discussed later in this chapter) in order to establish satisfactory representativeness.

Examples of Sampling Frames

The ultimate accuracy of a sample depends in large part on how well the sampling frame is constructed. This concept is so important that several examples of appropriate sampling frames follow:

• A survey of women of childbearing age is to be conducted. Since no such list exists, the researcher must operationalize this general population. This is likely to be accomplished by defining "childbearing age" in some reasonable manner (for example, fourteen to forty-five years of age) and then identifying women in this age bracket for the research. The sample could be drawn from such lists or screened by qualifying respondents through a set of introductory questions.

• A survey of people affected by noise generated by a local airport is to be implemented. The sampling frame can be established by identifying officially designated noise impact areas and obtaining a list of households within those boundaries. The sample would then be drawn from this list.

• A researcher is interested in criminal behavior and therefore plans to survey individuals who have committed a felony. Initially he or she obtains permission to interview inmates in jails and prisons. If this jail and prison population were used to represent the entirety of the sampling frame, systematic bias could be suspected because of the omission of criminals not currently incarcerated (those who either have been released or were never imprisoned in the first place). The researcher can address this suspected bias by including in the sample individuals currently on parole or probation to represent the released criminal population. The researcher may also be able to obtain a list of former convicts whose parole or probation has been successfully completed. However, the researcher cannot readily identify individuals who have committed felonies but have not been identified by the criminal justice system. The researcher can conclude that the absence of such individuals from the working population represents a systematic bias and can terminate the study, redefine the general population to include convicted criminals only, or proceed with the study with the appropriate disclaimer. In the alternative, however, it might not be unreasonable for the researcher to proceed without any disclaimer if it can be satisfactorily established that it is unlikely that this systematic omission is significant. In the example, the researcher may conclude that there is no major difference, in terms of identifying aspects of criminal behavior, between those who have been apprehended and those who have not.

Simple Random Sampling

There are several methods of drawing probability samples from the working population. The method chosen depends on a variety of factors, such as the manner in which the sampling frame is constructed and the focus of the study. The best-known form of probability sample is the *simple random sample*. The usual procedure is to assign a number to each potential respondent, or *sampling unit,* in the working population. Numbers are then chosen at random by a process that does not tend to favor certain numbers or patterns of numbers, and the sampling units selected become part of the sample itself. A common procedure for accomplishing this random process is to use computers to generate a random sample.

In a recent study of time demands on peace officers, it was determined that a random sample of 3,000 officer-days would be selected for the study. Randomization was accomplished by inputting the officers' station numbers, patrol unit number, shift, and date. Taken together, these four components constituted a full identification of an officer's daily patrol log that included a full record of time spent performing various police functions on any particular day. The computer was used to combine these components (for example, "4110A020404" refers to "station 41, unit 10, shift A, February 4, 2004") and select 3,000 at random. Telephone surveys use this process by randomly dialing telephone numbers—known as random digit dialing. In this way, the survey will reach all telephone numbers, including those that are unlisted.

In the absence of a computer for randomization, a table of random numbers can be used. Table 9.1 is an abridged table of random numbers. It is used in the following way. Suppose that there are 500 people in a working population and a researcher wishes to select a random sample of 10 persons. Each person must be assigned a number ranging from 001 to 500. Using Table 9.1, the researcher randomly selects a starting column or row of numbers and proceeds in any chosen direction (upward, downward, or across), looking for numbers between 001 and 500. To identify numbers from 001 to 500, the researcher must choose any three of the five digits given in each column. In this case, we assume that the process starts at line 9, column 2, and that it has been decided to proceed downward, looking at the first three digits in each random number for numbers between 001 and 500. The first number encountered is 975; the second is 633. Neither of these numbers falls within the required range, so the search is continued with 952, 669, and 131. The first sample member has been found: sampling unit number 131. The next relevant number is 341, at the bottom of column 2. The researcher then continues on to the top of column 3; proceeding in this way, the third number of the sample is located: 243. You should verify that the other seven sample numbers are as follows: 003, 056, 093, 305, 432, 497, and 311. If this process

TABLE 9.1. ABRIDGED TABLE OF RANDOM NUMBERS.

Line	Column				
	1	2	3	4	5
1	11404	10478	24317	60312	25164
2	65621	95574	93724	49741	65251
3	93998	73709	00325	78627	36815
4	22667	52883	05673	74698	64385
5	33362	68724	52681	31148	83761
6	07236	66537	70834	33260	72583
7	31768	30247	90313	77538	05367
8	54121	21768	09324	79572	29734
9	68417	97521	56698	09525	76354
10	93561	63399	84743	39751	29448
11	31790	95267	75464	05783	98523
12	48585	66947	30541	64728	90400
13	93614	13143	58366	05070	37304
14	00071	86770	43287	07386	16458
15	48277	34132	73045	41818	07465

causes the researcher to arrive at a number that has already been selected for the sample, it is simply skipped, consistent with the premise of sampling without replacement.

Systematic Random Sampling

Systematic random sampling is an adaptation of the simple random process used when the working population list is quite large and the sampling units cannot be conveniently or feasibly numbered. If the working population consisted of 3 million people and the drawing of a sample of 1,500 people was required, the process of numbering and selecting from a table of random numbers would be prohibitively burdensome, as would the inputting of 3 million cases simply in order to select 1,500 of them.

A systematic sample assumes that the working population list is randomly distributed; therefore, the researcher can systematically choose sample members by selecting them from the list at fixed intervals (every nth entry). For instance, the 1,500 sample members represent 1 out of 2,000 people on the working population list (3,000,000/1,500 = 2,000), so if the selection process were to count to every 2,000th sampling unit from a starting point on the list selected at random from the first 2,000 cases, it would be expected that a random sample would result. If the starting point were selected outside of the first 2,000 cases, the list could be exhausted before 1,500 respondents were chosen. Once again, the table of random numbers can be useful in the sample selection process—this time for choosing a starting point. Without looking at the table, the selector should arbitrarily pick a starting point (for example, column 4, line 15). Better yet, the selector can eliminate even this small potential for bias by asking someone else to choose a column and row or to point to a position on the table while blindfolded.

It is rare for the systematic process to yield fixed intervals that are whole numbers without a fractional remainder. In the example, if the working population were 3,100,000, the requisite fixed interval would be 2,066.67, which would require truncating the decimal. As an example, suppose that a working population consists of a list of 250 students and that the researcher requires that a sample of 9 students be selected. Dividing the working population of 250 by the desired sample size of 9 yields 27.8. This decimal is truncated (not rounded), leaving the whole number 27; hence, every twenty-seventh person on the list would be selected. The starting point would be any number between 1 and 27, selected randomly. If the starting point happens to be 7, then the sample of 9 would consist of the following sampling units: 7, 34, 61, 88, 115, 142, 169, 196, and 223. If the decimal were rounded instead of truncated, the sample would include student number 252, which does not exist in the population of 250 students. Truncation does tend to eliminate certain sampling units from the possibility of being selected. In this case, students 244 through 250 would not have a chance of being selected. However, since the working population list is assumed to be random, this tendency to eliminate a small number of possible sample members is also random, and therefore it is far more acceptable than choosing nonexistent sampling units.

Taking this process one step further, let it be supposed that the persons numbered 34, 61, and 115 refuse to respond or cannot be reached for an interview. The researcher is then faced with having to elect three alternate respondents. Systematic sampling proceeds as follows. The remaining unselected working population sampling units are renumbered from 1 to 241 to account for the 9 sampling units previously selected. This remaining number is divided by the required three respondents. Thus, 241 is divided by 3 to yield 80.3. Then, every 80th person is selected, starting with any number (selected at random) between 1 and 80. If the starting number is 60, the sample of three would include the 60th, 140th, and 220th person on the list. This procedure is repeated, if necessary, until all respondents have been successfully interviewed.

Cluster (Multistage) Sampling

Another variation on simple random sampling is known as *cluster* or *multistage sampling*. A cluster sample is a sample in which there is a hierarchy of sampling units.

The primary sampling unit is a grouping (or cluster) of the individual elements that are the focus of the study. This grouping must be a well-delineated subset of the general population that is considered to include characteristics found in that population. Such groupings typically consist of counties, cities, census tracts, census blocks, and so forth. A random sample of these units is selected. Secondarily, a subset of smaller units within the primary units is randomly selected. This process continues at various stages until the researcher has selected a random sample of the actual units of analysis. Cluster sampling arises predominantly from situations in which the population is so prohibitively cumbersome that traditional random sampling techniques cannot be easily employed.

To illustrate a situation in which cluster sampling can be used, suppose that a mail survey of 1,000 residents of the United States is to be conducted. Traditional sampling methods would require that a list of U.S. addresses be available. Such a list would be very difficult to obtain and, at the very least, very cumbersome to process. Cluster sampling can provide a multistage procedure that will alleviate this problem. This procedure might entail randomly selecting counties, then census tracts within the counties, and finally individual households in those census tracts.

In certain situations, clusters are known to be substantially different from one another in terms of size or homogeneity of their populations. For instance, in most states, the population tends to be concentrated in a handful of counties. If clustering is to be performed by county, it is possible to underrepresent or even completely overlook these large counties. To avoid this lack of representativeness, it is necessary to establish a list of large counties and smaller counties based on population size (for example, 1 million residents and above versus fewer than 1 million residents). Clusters can then be selected from each of these groups to ensure adequate representation.[1]

A further illustration of cluster sampling involves an onboard survey of bus riders in a major metropolitan county that we have performed for several transit agencies. Inasmuch as there is no efficient way to obtain a comprehensive list of bus riders, it is necessary to consider a cluster sampling procedure, as in this example of a 1,200 rider sample:

1. List all bus routes in the system by direction of travel, day of the week, and time of day. This list will take the form of a matrix of route-and-time cells containing one bus route for a specific direction, at a certain time of day, on a particular day of the week.
2. Each cell is weighted according to the known average volume of bus passengers. These weighted cells constitute the primary sampling frame for sample selection.

3. Select a random sample of 400 weighted cells. The selected cells indicate the buses earmarked for onboard interviews.
4. To obtain the 1,200 personal interviews, 3 randomly selected passengers on each of the 400 selected buses are surveyed.

Stratified Sampling

The example of selecting clusters from both large and small counties leads to a discussion of stratified sampling. *Stratified sampling* consists of separating the elements of the working population into mutually exclusive groups called *strata;* random samples are then taken from each stratum. For example, a researcher may be interested in voter opinion concerning the issue of gun control. It is considered important to analyze the population by ethnic group. Accordingly, the sampling frame is separated into strata based on ethnicity. The primary purpose in this sample selection process is to make certain that each stratum is represented by an adequate sample size in order to analyze the stratum both separately and as a part of the total population. This is much more likely to occur when selection is performed by stratum than by an overall population random sampling procedure. Table 9.2 presents a hypothetical overall breakdown of the working population by ethnic group and the expected random sample representation of the various groups proportionate to their number in the overall population, based on a sample size of 600.

If the researcher were to proceed with random sampling procedures, it could be expected that the sample sizes of the ethnic strata would approximate the proportions they represent in the overall population. Hence, the researcher might consider, in advance of ultimate sample selection, that the sample is likely to have approximately 360 whites, 90 blacks, 90 Hispanics, and 60 Asians. However, since ethnicity has been established as an important criterion in the study, the researcher must now recognize that the number of blacks, Hispanics, and Asians to be

TABLE 9.2. PROPORTIONATE SAMPLE REPRESENTATION FOR A HYPOTHETICAL ETHNIC DISTRIBUTION.

Ethnic Group	Population Size (in thousands)	%	Expected Proportionate Sample Representation	%
White	6,000	60.0	360	60.0
Black	1,500	15.0	90	15.0
Hispanic	1,500	15.0	90	15.0
Asian	1,000	10.0	60	10.0
Total	10,000	100.0	600	100.0

sampled in the study is probably going to be too small to achieve certain requisite margins of error. Chapter Eight indicates that these expected group sizes do not meet sample size requirements for even a margin of error of ±10 percent (95 percent level of confidence), thereby calling into question the researcher's ability to make reasonably accurate generalizations concerning these groups. A practical rule of thumb is that a 10 percent margin of error is the maximum that should be tolerated for any group or subgroup within the overall sample. Hence, a sample size of approximately 100 is required for all strata and substrata. Therefore, if the researcher wants to attempt to achieve at least the ±10 percent margin of error for each stratum, the overall sample size will have to be increased accordingly to reasonably ensure that each group will meet that threshold. For example, since Asians are the smallest ethnic group in terms of numbers and they represent 10 percent of the population, a total sample size of a minimum of 1,000 persons will be required to anticipate approximately 100 Asian respondents within that sample. Increasing the sample size can have serious cost considerations, however, and may not be feasible within the researcher's budget or time frame.

Stratified sampling offers to the researcher a method by which the margin-of-error requirement of a maximum of ±10 percent for each stratum can be satisfied while still keeping the overall sample size at 600, as long as at least 100 persons are interviewed in each stratum. The recommended disproportionate sample sizes by stratum for this particular example are:

White	300
African American/black	100
Hispanic/Latino	100
Asian	<u>100</u>
Total	600

This disproportionate sampling distribution is obtained by expanding the sample size of strata that would otherwise not attain the requisite minimum size and correspondingly reducing the group or groups that would achieve the requisite minimum through the standard random sampling process. The samples must now be selected randomly, at least to some partial extent, from four separate sampling frames within the total population. At this point, in the case of small populations for any stratum, equation 8.7 and Table 8.2 should be used to determine the sample size of each stratum. This will reduce the requisite sample sizes.

The difficulty with disproportionate stratification is that the overall sample is now skewed toward the smaller strata. In the case at hand, African Americans, Latinos, and Asians are overrepresented, and whites are underrepresented. A weighting procedure must be employed to analyze the data with regard to the total population.

Weighting Disproportionate Samples

In situations where the survey has produced disproportionate results—either deliberately stratified or due to systematic omissions that are correctable by weighting—an adjustment is required to make certain that the total sample is a proportionate representation of the population rather than a summation of disproportionately sampled groups within that population. This adjusted total sample is to be used when analyzing the total population, whereas the individual disproportionate strata will be used for the analyses for each of those subgroups.

This corrective weighting process is best demonstrated by an example of disproportionate sampling in the City of Davis, California. For planning purposes, Davis and its environs are divided into a series of planning areas, eight of which were targeted for study. These planning areas are listed in Table 9.3, along with their populations and corresponding percentages.

It was determined that the importance of these planning areas was sufficiently significant to require that each planning area be separately analyzed in accordance with a maximum margin of error of ±10 percent (95 percent confidence level). Budget and time considerations limited the overall sample size to approximately 800 persons. In large populations (100,000 or more), this margin of error is achievable with a sample size of approximately 100 persons. For smaller populations, the requirement is reduced somewhat (see Chapter Eight). Therefore, appropriate sample sizes were obtained to fulfill this objective; these are indicated in Exhibit 9.1 as "Actual Sample Size."

It was equally important in this study to analyze the results on a citywide basis. This required that the sample from each planning area be weighted to reflect that area's relative population within the city. Exhibit 9.1 demonstrates the calculation

TABLE 9.3. POPULATION OF THE DAVIS PLANNING AREAS.

Planning Area	Population	%
North Central	2,026	3.7
West Davis	8,510	15.4
Central	19,971	36.1
East Davis	13,360	24.1
East Davis–Mace	939	1.7
Core	1,258	2.3
South Davis	8,268	14.9
South Davis–County	997	1.8
Total	55,329	100.0

EXHIBIT 9.1. SAMPLE SIZES AND WEIGHTS FOR THE CITY OF DAVIS PLANNING AREAS.

Planning Area	Expected Proportionate Sample Size	Actual Sample Size	Weight (Expected/Actual)
North Central	31	103	.30
West Davis	128	106	1.21
Central	301	110	2.74
East Davis	201	110	1.83
East Davis–Mace	14	104	.13
Core	19	105	.18
South Davis	124	103	1.20
South Davis–County	15	92	.16
Total	833	833	

of the weights to be applied to the data from each planning area to generate a citywide total. The calculation of these weights entails identifying the expected number of survey respondents in each planning area if the survey had been conducted randomly without stratification. Hence, for example, since the Central area represents 36.1 percent of the population of Davis, it is expected that approximately 36.1 percent of 833 randomly selected citywide participants (301 persons) would be residents of the Central planning area in that survey. The expected proportionate sample size for each area is then compared to the sample size actually obtained, and a weight is calculated by dividing the expected proportionate sample size by the actual sample size.

Table 9.4 depicts the outdoor recreation facilities used most often by Davis residents, cross-tabulated by planning area. The column totals for each individual planning area represent the actual sample data, and the "Weighted Citywide Total" column reflects the application of the weights from Exhibit 9.1. This calculation is depicted in Worksheet 9.1, which selects categories of the recreational facility variable and applies the weight for each planning area to the actual sample data for those categories. The worksheet then sums these weighted data to obtain a weighted citywide total, which is rounded to whole numbers to reflect the discrete nature of the data: individual respondents. The process continues through each category of the dependent variable. Worksheet 9.1 depicts this process for two categories of the dependent variable. You can verify that this process has been applied to the other eight categories in Table 9.4.

TABLE 9.4.   OUTDOOR RECREATION FACILITIES USED MOST OFTEN BY DAVIS RESIDENTS, BY PLANNING AREA.

Planning Area

Outdoor Facility	North Central		West		Central		East Davis		East Davis–Mace		Core		South Davis		South Davis–County		Weighted Citywide Total[a]	
	f	%	f	%	f	%	f	%	f	%	f	%	f	%	f	%	f	%
Greenbelts	56	28.9	31	16.6	45	26.8	35	18.8	54	31.4	26	14.8	28	17.2	10	9.6	288	21.3
Athletic fields	34	17.5	39	21.0	26	15.5	27	14.5	23	13.4	30	17.0	24	14.7	19	18.3	218	16.1
Children's play areas	26	13.4	19	10.2	18	10.7	33	17.8	24	13.9	15	8.5	22	13.5	18	17.3	175	13.0
Lawn areas	23	11.8	14	7.5	21	12.5	14	7.5	13	7.5	32	18.2	15	9.2	8	7.7	134	9.9
Swimming pools	12	6.2	26	14.0	9	5.3	24	12.9	19	11.0	16	9.1	15	9.2	8	7.7	128	9.5
Picnic areas	10	5.2	7	3.8	8	4.8	18	9.7	18	10.5	15	8.5	17	10.4	13	12.5	94	7.0
Tennis courts	14	7.2	17	9.1	12	7.1	8	4.3	8	4.7	6	3.4	10	6.1	7	6.7	88	6.5
Basketball courts	9	4.6	17	9.1	10	6.0	9	4.8	8	4.7	16	9.1	6	3.7	7	6.7	79	5.9
Other[b]	10	5.2	12	6.5	10	6.0	12	6.5	5	2.9	16	9.1	20	12.3	12	11.6	97	7.2
None used	0	0.0	4	2.2	9	5.3	6	3.2	0	0.0	4	2.3	6	3.7	2	1.9	49	3.6
Total[c]	194	100.0	186	100.0	168	100.0	186	100.0	172	100.0	176	100.0	163	100.0	104	100.0	1,350	100.0

[a]Because of the nature of the weighting process, the sum of the planning area frequencies does not equal the weighted citywide total.

[b]"Other" includes such outdoor recreation facilities as skate parks, golf courses, parks, the arboretum, bicycle trails, and community gardens.

[c]Because respondents were provided the opportunity to give more than one response, column totals may exceed sample size.

WORKSHEET 9.1. CALCULATION OF WEIGHTED CITYWIDE TOTAL FOR OUTDOOR RECREATION FACILITIES USED MOST OFTEN.

		Facility	
Planning Area	**Weight**	**Greenbelts**	**Athletic Fields**[a]
North Central	0.30	56 × 0.30 = 16.80	34 × 0.30 = 10.20
West Davis	1.21	31 × 1.21 = 37.51	39 × 1.21 = 47.19
Central	2.74	45 × 2.74 = 123.30	26 × 2.74 = 71.24
East Davis	1.83	35 × 1.83 = 64.05	27 × 1.83 = 49.41
East Davis–Mace	0.13	54 × 0.13 = 7.02	23 × 0.13 = 2.94
Core	0.18	26 × 0.18 = 4.68	30 × 0.18 = 5.40
South Davis	1.20	28 × 1.20 = 33.60	24 × 1.20 = 28.80
South Davis–County	0.16	10 × 0.16 = 1.60	19 × 0.16 = 3.04
Weighted citywide total[b]		288	218

[a]Continue same procedure for each outdoor recreation facility.

[b]Summation of weighted planning area totals.

EXHIBIT 9.2. SAMPLE SIZES AND WEIGHTS FOR STRATIFIED SAMPLE FOR TABLE 9.2.

Ethnic Group	Expected Proportionate Sample Size	Actual Sample Size	Weight (Expected/Actual)
White	364	304	1.197
African-American/Black	91	100	.910
Latin/Hispanic	91	102	.892
Asian	61	101	.604
Total	607	607	

As another example, suppose that the survey that was proposed in Table 9.2 was performed and the ultimate distribution of responses was as follows:

White	304
African-American	100
Hispanic/Latino	102
Asian	101
	607

Exhibit 9.2 shows the calculation of weights for this sample survey.

Nonprobability Sampling

The essential characteristic of nonprobability sampling is that the researcher does not know the probability that a particular respondent will be selected as part of the sample. Therefore, there is no certainty that the probability of selection is equal among the potential respondents. Without such equality, the researcher cannot analyze the sample in the context of the normal distribution. Therefore, the sample data cannot be used to generalize beyond the sample itself, because the degree of sampling error associated with the sample cannot be estimated without an assumption of normality. Intercept surveys in particular are likely to be nonprobability without significant preparatory work and careful planning to ensure representativeness.

Nonprobability samples do not provide the researcher with the ability to generalize survey data with a known degree of accuracy. That is, the information obtained, although useful, does not allow a specific margin of error to be identified, which interferes with any attempt to scientifically generalize the findings.

In spite of this shortcoming, nonprobability sampling can be helpful to the researcher. It is considerably less complicated in terms of strict adherence to the tenets of random sample selection and is therefore much less costly and time-consuming than probability sampling. The primary advantage of nonprobability sampling rests in its usefulness in the preliminary stages of a research project or in a postsurvey debriefing phase to elaborate on certain survey findings that are particularly difficult to explain. In Stage 3 of the survey process (see Chapter One), for instance, the researcher must ensure that there is adequate knowledge of the investigative area before constructing specific questions. The use of a nonprobability sample can quickly generate a preliminary understanding of some of the key issues underlying the research study. It is also the primary means by which researchers pretest and refine their survey instrument, as discussed in Chapter Two.

The most common example of a nonprobability sample is a *sidewalk survey*, where interviewers interview passersby at, for example, a shopping center. The general population in this example is shoppers. Strict adherence to the principles of probability sampling would require the compilation of a count of all such shoppers and the ability to access all shoppers as the working population. In the case of a sidewalk survey, this working population is typically not enumerated; consequently, the probability of selecting any particular passerby from that working population cannot be determined. Furthermore, under these circumstances, there can be a significant element of individual interviewer discretion in the selection of interviewees, which might compound the existing uncertainty about whether the sample truly represents the general population.

There are several types of nonprobability samples. The sidewalk survey is an example of convenience sampling, in which interviewees are selected based on their presumed resemblance to the working population and their ready availability. Frequently students are interviewed in their classrooms, enabling the researcher to contact large numbers of respondents in a relatively short period of time and at minimal cost. It is important to reemphasize that the researcher cannot generalize the findings in such cases beyond the sample itself. These findings can be used only as an informal base of knowledge, in preparation for a survey research project based on probability sampling or for purposes of elaborating on the otherwise undetected nuances, themes, and patterns of the population already quantitatively surveyed. Note that convenience sampling and *snowball sampling* are at the core of obtaining participants for focus groups (see Chapter Four).

Snowball sampling is particularly beneficial when it is difficult to identify potential respondents. Once a few respondents are identified and interviewed, they are asked to identify others who might qualify as respondents. Suppose that a researcher has initiated a study that requires interviewing drug abusers who have not sought medical or social assistance. Quite obviously such respondents could not be easily identified. However, the researcher may be able to identify and interview a small number of drug abusers using personal reconnaissance. Snowball sampling could then be invoked by relying on these initial respondents to provide access to other drug users. Another type of nonprobability sample is known as the purposive sample. In the purposive sample, the researcher uses his or her professional judgment, instead of randomness, in selecting respondents. For example, a researcher may be interested in gathering information about problems related to juveniles in a particular community. *Key respondents,* whom the researcher considers to be particularly knowledgeable about the subject, may be selected for interviews. These respondents may include such people as the directors of social service agencies, law enforcement personnel, judges, attorneys, and educators. Responses to a set of questions may then be summarized as part of a larger study concerning juvenile problems.

EXERCISES

1. You are interested in performing a survey that involves an attempt to identify the reasons associated with vacation preferences among tourists in New Orleans, Louisiana.
    a. Suggest possible informational sources for identifying a working population.
    b. Indicate an appropriate working population from which the sampling frame can be constructed.

    c.   What systematic biases might exist in your sampling frame?

    d.   Can the cluster sampling technique be used to determine a sample? If so, how?

2.   Twenty-five persons out of a class of 75 students are to be randomly selected for participation in a certain experimental project. Use Table 9.1 to select the 25 participants.

3.   You wish to survey 600 health professionals regarding their perception of medical ethics. The American Medical Association has provided an unnumbered list of the 1,432,000 health professionals in your region of the United States. Apply the procedure of systematic random sampling to indicate where on this list you would start the sample and how you would proceed to obtain the complete sample.

4.   The list provided by the American Medical Association in Exercise 3 also indicates the ethnic background of each professional. The percentage ethnic breakdown is as follows:

Ethnicity	%
White	61.7
Asian	14.8
Hispanic	13.6
Black	9.9
	100.0

    The study seeks to make generalizations according to ethnicity with a maximum margin of error of ± 10 percent for each ethnic group.

    a.   What would you expect to be the ethnic distribution of your 600–person sample in Exercise 3?

    b.   You determine that a disproportionate stratified sample is needed to satisfy these requirements. What should be the actual stratified ethnic breakdown of your sample to satisfy the required margins of error?

    c.   If your ultimate sample exactly mirrors your proposed stratified sample, what weights would you apply to the data in your analysis?

5.   For each of the following situations, briefly explain whether (and why) the sample selected is representative of the general population indicated:

    a.   A report indicated that 175 University of Notre Dame students (100 women and 75 men) were randomly selected from students dining at the University Commons on a Tuesday afternoon in November. The study was to identify eating habits of the current student body.

    b.   It was reported that data were obtained for a stratified random sample of persons aged sixty and over living in the community of Munderotz, New Jersey. The final sample contained 301 persons married and living with their spouse, 423 who lived alone, and 560 who lived with someone other than their spouse. The study sought to determine living arrangements for senior citizens in New Jersey.

    c.   Researchers studied 910 children who were pupils in public schools in Norway. The country was divided into seven equally populated sectors, with ten schools randomly chosen from each sector. Thirteen children were then randomly selected from each school. The study undertook to assess the academic performance of public school pupils in Norway.

6. A study of undocumented immigrants is about to be undertaken. The researchers find that they do not possess information about the important issues facing these individuals sufficient to construct a thorough survey questionnaire. However, to further their research interests, researchers decide to gather preliminary information from this group. How can the researchers use nonprobabilistic sampling techniques to obtain this preliminary information?

    a.   Convenience sampling.

    b.   Purposive sampling.

    c.   Snowball sampling.

7. A city with a population of 180,000 is the subject of a sample survey. According to the most recent census, the city's political party breakdown is as follows:

Democrat   125,000

Republican   30,000

Independent   25,000

    A survey of 500 persons has been commissioned from you.

    a.   How many residents would you expect from each group based solely on proportional representation?

    b.   How would you stratify the total sample so as to be able to have a maximum of a ±10 percent margin of error for each group? (Note small sample sizes where appropriate.)

    c.   What weights would you assign to each group in order to combine them into an overall citywide total if the sample ultimately contained 310 Democrats, 101 Republicans, and 100 Independents?

    d.   What is the approximate margin of error (overall) for this survey?

Note

1. The most extensive treatment of a method of selecting the number of clusters to be sampled is contained in Schaeffer, Mendenhall, and Ott (1986). They present a complex and innovative formula for determining the number of clusters to be included in a cluster sample, which can be shown at the 95 percent level of confidence, as follows:

$$n = \frac{N\sigma_c^2}{N(B^2\bar{M}^2/4) + \sigma_c^2}$$

where

    n = number of clusters selected in a simple random sample

    N = number of clusters in the whole population

σ_c^2 = population variance associated with the sizes of the clusters in the population
$\quad B$ = margin of error (confidence interval) in terms of either proportions or interval data
$\overline{M}$ = mean cluster size of the whole population

The formula will tend to yield a number of clusters to be sampled that is somewhat lower than the number generated by Tables 8.1 and 8.2. It does so because of two factors:

a. A preliminary sample is required to estimate $\overline{M}$ and σ_c^2.
b. There is an implicit assumption that there will be a full canvassing of all members of the clusters selected.

Hence this formula has the advantage of a greater geographical concentration of clusters, but there are disadvantages in terms of the costs of conducting an adequate preliminary sample and obtaining a 100 percent census within the clusters themselves. Also, because of the smaller number of clusters, there is a greater likelihood of needing to stratify the cluster sample. On balance, we believe it to be more consistent with the needs of the readers of this book, in terms of practical applications, cost factors, and other such considerations, to use the approaches presented in Chapters Eight and Nine in the application of cluster sampling.

PART THREE

PRESENTING AND ANALYZING SURVEY RESULTS

CHAPTER TEN

ANALYZING CROSS-TABULATED DATA

The distribution of a single variable is described by the frequency distribution, and this univariate relationship is further clarified and understood through the use of measures of central tendency and dispersion. Multivariate relationships (those with more than one variable) can be presented through the use of contingency tables, but unlike frequency distributions, descriptive statistics are not adequate to analyze such relationships. Inasmuch as the primary purpose of contingency tables (cross-tabulations) is to depict the relationship between two or more variables, tests of statistical significance and measures of association (analytical statistics) need to be applied to the data to verify the existence and strength of any apparent relationships between variables.

This chapter presents the most commonly applied tests of statistical significance and measures of association that relate particularly to cross-tabulated survey data at the nominal or ordinal level of measurement. These analytical methods are essential for researchers to understand because they have the ability to discriminate among voluminous amounts of data generated by the survey research process. They permit researchers to identify relationships among survey variables and address whether data from a sample can be used to represent facts about the general population from which the sample has been drawn.

Chapters Eleven and Twelve continue the discussion of analytical statistics, presenting several additional tests of significance and measures of association that are particularly useful in the analysis of interval level sample survey data. Chapter Thirteen discusses the presentation of the analyzed findings in a final report.

Cross-Tabulated Contingency Tables

The most elementary tabular display of data is the frequency distribution. A *frequency distribution* is a summary presentation of the frequency of response (f) of each category of a variable. An example of a frequency distribution is provided in Table 10.1. The frequency distribution provides the response frequencies for each question in a survey.

Frequency distributions involve a description of one variable. Often, however, the research calls for a simultaneous analysis of more than one variable. For example, a researcher may be interested in knowing the relationship between the ethnic background of the survey respondents and their educational attainment in order to recommend certain culturally based proposals for increasing educational attainment in a study area. In such a case, two variables are under study: ethnicity and education.

Since the researcher is interested in the influence one variable may have on another, the use of a contingency table is appropriate. Contingency tables add an explanatory dimension to the frequency distribution. Table 10.2 is an example of a contingency table drawn from the same data as Table 10.1 but including two variables: years on the reservation and tribe.

If the data contained in Table 10.2 had been presented in frequency distributions, the researcher would have found that the American Indians surveyed were relatively evenly split among the years on the reservation groupings and that there were more Pala surveyed than Rincon and San Pasqual, as indicated in the "Total" column and row. The contingency table provides the additional information

TABLE 10.1. NUMBER OF YEARS RESIDED ON THE RESERVATION.

Number of Years	f	%
Less than 1	101	13.1
1 and under 5	145	18.7
5 and under 10	149	19.3
10 and under 20	135	17.4
20 and under 30	134	17.3
30 and more	110	14.2
Total	774	100.0
Nonresponses = 26		

TABLE 10.2 NUMBER OF YEARS RESIDED ON THE
RESERVATION BY TRIBE.

| | Tribe | | | | | | | |
| | Pala | | Rincon | | San Pasqual | | Total | |
Years	f	%	f	%	f	%	f	%
Less than 1	49	13.1	38	17.3	14	7.8	101	13.1
1 and less than 5	77	20.6	31	14.1	37	20.7	145	18.7
5 and less than 10	75	20.0	46	20.9	28	15.6	149	19.3
10 and less than 20	62	16.5	35	16.9	38	21.2	135	17.4
20 and less than 30	71	18.9	38	17.3	25	14.0	134	17.3
30 or more	41	10.9	32	14.5	37	20.7	110	14.2
Total	375	100.0	220	100.0	179	100.0	774	100.0

that the particular tribes differ in length of time on their reservation. These data are derived by a computer-generated cross-tabulation of two questions from the survey instrument and their associated variables: years on the reservation and tribe. Table 10.2 reveals an apparent difference among these groups that reflects the San Pasqual tribe's being somewhat longer tenured (20.7 percent thirty or more years and 7.8 percent less than one year) than the other two.

Important principles apply specifically to contingency tables. Prior to enumerating these principles, it is critical to explain the concepts of independent and dependent variables.

The *independent variable* is the change agent, or the variable that attempts to explain changes in the dependent variable. It acts on, influences, or precedes the *dependent variable,* which is the variable that is being explained and is therefore dependent on the independent variable. In the case of the variables in Table 10.2, the tribe of the respondent is the independent variable that is said to have some influence on years on the reservation. Reversing these characterizations is logically much less likely.

In determining which variable should be considered independent and which should be considered dependent, the researcher can use a certain rule of thumb if common sense does not clearly specify an independent variable for a given situation. An attempt should be made to identify which variable came first in time, and that variable can be specified as independent. For example, in Table 10.2, tribe clearly precedes length of time on the reservation temporally; in the case of two variables, such as "years of schooling" and "parents' socioeconomic class," that logic would dictate that parents' socioeconomic class be designated as independent because it too precedes the children's educational achievement.

When a temporal relationship is not clear, the researcher must use professional judgment about the study's intent and focus to choose which variable should be treated as the independent variable. A statistical measure is available to help in this determination. This measure is called lambda (λ).[1]

Contingency tables include a "Total" column in addition to the "Total" row that is shown in typical frequency distributions. The Total column depicts the sum of all categories of the column variable for each category of the row variable. Inasmuch as the stated purpose of a contingency table is to determine the relationship between two variables, the inclusion of a Total column is a critical and necessary component. In addition, the table should be prepared so that the independent variable is the column variable—the one along the top—and the dependent variable is the row variable. The title of the table should be expressed in terms of the dependent variable "by" the independent variable.

Percentages are calculated for the independent (column) variable only. In accordance with this principle, percentage is calculated vertically, summing to 100 percent at the bottom of each column.[2] The variables can be compared by holding a category of the dependent variable (row) constant and comparing the percentages across the row. For example, in Table 10.2, whereas 16.5 percent of Pala tribe members have lived on the reservation for ten and less than twenty years, 21.2 percent of San Pasqual members have lived on the reservation that long.

At least two situations may arise that would prompt the researcher to reverse the independent and dependent variables. First, the number of categories of the independent variable may be so extensive that it is difficult to fit them on a single sheet of paper for final report purposes. In such a case, it may be necessary to present the independent variable vertically in order to reconfigure the table so that it can be contained on one page. Second, it is possible that in a series of tables, one variable will be consistently identified as either independent or dependent (for example, income is generally an independent variable in opinion and behavioral surveys), but an isolated instance may arise that identifies the variable in the opposite way (for example, income cross-tabulated with ethnicity). For this single situation, the researcher may decide not to reverse the axis to which the reader has become oriented. In cases where the independent variable is located on the rows rather than across the top, percentages should still be calculated for the independent variable (although this will occur horizontally), and comparisons between the variables will be accomplished instead by comparing percentages down the columns.

It is possible to add a third variable to a contingency table analysis by constructing contingency tables that cross-tabulate the dependent variable and the independent variable while holding each category of a third (control) variable

TABLE 10.3.   NUMBER OF YEARS RESIDED ON THE RESERVATION, BY TRIBE (MALE RESPONDENTS).

| | Tribe | | | | | | | |
| | Pala | | Rincon | | San Pasqual | | Total | |
Years	f	%	f	%	f	%	f	%
Less than 1	26	13.0	20	16.7	5	5.5	51	12.4
1 and less than 5	40	20.0	16	13.3	15	16.7	71	17.3
5 and less than 10	30	15.0	27	22.5	15	16.7	72	17.6
10 and less than 20	44	22.0	20	16.7	24	26.7	88	21.5
20 and less than 30	40	20.0	22	18.3	15	16.7	77	18.8
30 or more	20	10.0	15	12.5	16	17.7	51	12.4
Total	200	100.0	120	100.0	90	100.0	410	100.0

constant. For example, Table 10.2 could be presented in a more detailed format by preparing a series of contingency tables that display tenure on the reservation by tribe and gender. Table 10.3 is an example of one of these gender-based distributions. This is ordinarily referred to as a three-way cross-tabulation; it is particularly useful when it is suspected that additional variables are involved in the relationship. Men seem to have lived on the reservation slightly longer than have women. This is found in that men have lower percentages in Table 10.3 than the overall (Table 10.1) for the shorter periods and higher percentages for the longer tenures, with the exception of the longest period, which is likely due to female longevity.

The Chi-Square Test of Significance

From a cursory reading of Table 10.4, one might make the following assumptions: Democrats tend to favor gun control more than either Republicans or Independents (56.5 percent versus 52.4 percent and 50.0 percent, respectively), and Independents tend to have no opinion on the issue of gun control more than either Democrats or Republicans (16.7 percent versus 10.9 percent and 11.9 percent, respectively).

These perceived differences may actually exist within the general population, or they may simply be the result of built-in sampling error in the random sampling process. Determination of the statistical validity of these perceived differences is made through statistical significance tests, of which the most frequently used in survey research is the chi-square test. Chi square is the only significance test available for data with both variables measured on the nominal scale. However,

TABLE 10.4. OPINION CONCERNING GUN CONTROL BY POLITICAL PARTY.

| | Political Party | | | | | | | |
| | Democrat | | Republican | | Independent | | Total | |
Opinion	f	%	f	%	f	%	f	%
Favor	130	56.5	110	52.4	30	50.0	270	54.0
Do not favor	75	32.6	75	35.7	20	33.3	170	34.0
No opinion	25	10.9	25	11.9	10	16.7	60	12.0
Total	230	100.0	210	100.0	60	100.0	500	100.0

data measured on the ordinal and interval scale, organized into categories and presented in a contingency table, can also be tested using chi square.

The *chi-square test* of significance is essentially concerned with the differences between the frequencies that are obtained from the sample survey and those that could be expected to be obtained if there were no differences among the categories of the variables. As discussed in Chapter Seven, the assumption that no difference exists among the categories of the variables is known as the *null hypothesis.* The chi-square test seeks to identify whether the perceived findings are genuine and therefore generalizable to the full population or the result of sampling error, in which case they would not be generalizable. With reference to Table 10.4, under the assumption of no difference, the researcher would expect that the overall percentage of those who favor gun control (54.0 percent) would also be the percentage of Democrats, Republicans, and Independents who favor gun control. The essence of this assumption is that if the percentages associated with the categories are the same as those associated with the entire distribution, it can be said that the two variables in the contingency table have no relationship to each other; under these circumstances, political party affiliation would have no bearing on opinion concerning gun control, and any differences obtained by the sample would have occurred by sampling error alone. The chi-square test seeks to identify whether any differences among the categories of the variables in the sample are genuine or merely the result of sampling error.

Calculation of the chi-square statistic (χ^2) involves measuring the difference between the expected frequencies that are consistent with the null hypothesis and those frequencies obtained through the survey process, in accordance with the following equation:

$$\chi^2 = \Sigma \frac{(f_o - f_e)^2}{f_e} \tag{10.1}$$

WORKSHEET 10.1.   CHI-SQUARE MATRIX ($n = 500$).

	Democrat		Republican		Independent		Row Totals
	f_o	f_e	f_o	f_e	f_o	f_e	
Favor	130	(124.2)	110	(113.4)	30	(32.4)	270
Do not favor	75	(78.2)	75	(71.4)	20	(20.4)	170
No opinion	25	(27.6)	25	(25.2)	10	(7.2)	60
Column Totals	230		210		60		500

where f_o = the frequency obtained in each cell and f_e = the frequency expected in each cell under the assumption of no difference.

The first step in the calculation of chi square is the establishment of a chi-square matrix worksheet consisting of obtained and expected frequencies. Worksheet 10.1 contains the matrix for the data in Table 10.1. The numbers in each cell represent the obtained frequency (without parentheses) and the expected frequency (within parentheses) in which the overall percentage distribution of opinion is assumed, under conditions of no difference, to be replicated for each category of political affiliation. Thus, for each political affiliation, the expected frequencies reflect a distribution of 54.0 percent in favor, 34.0 percent not in favor, and 12.0 percent with no opinion.

An alternative method for calculating the expected frequencies is as follows. For each cell, multiply the row total corresponding to that cell by the column total and then divide the product by n. In the case of Democrats in favor of gun control, for example, the expected frequency of 124.2 can be calculated by multiplying 54.0 percent by the total number of Democrats ($0.54 \times 230 = 124.2$) or by multiplying 270 by 230 and dividing by 500 ($270 \times 230/500 = 62,100/500 = 124.2$).

Application of the chi-square formula yields the following calculation:

$$\chi^2 = \Sigma \frac{(f_o - f_e)^2}{f_e}$$

$$= \frac{(130 - 124.2)^2}{124.2} + \frac{(110 - 113.4)^2}{113.4} + \frac{(30 - 32.4)^2}{32.4} + \frac{(75 - 78.2)^2}{78.2}$$

$$+ \frac{(75 - 71.4)^2}{71.4} + \frac{(20 - 20.4)^2}{20.4} + \frac{(25 - 27.6)^2}{27.6} + \frac{(25 - 25.2)^2}{25.2} + \frac{(10 - 7.2)^2}{7.2}$$

$$= 0.271 + 0.102 + 0.178 + 0.131 + 0.180 + 0.008 + 0.241 + 0.002 + 1.120$$

$$= 2.23$$

To interpret the calculated chi-square, the researcher must refer to a table of critical chi-square values (see Exhibit 10.1). This exhibit shows the required magnitude of the calculated chi square in order to achieve statistical significance—or the rejection of the null hypothesis. That is, if the calculated chi square (χ^2) equals or is greater than the critical chi square from Exhibit 10.1 (χ^{2*}), the differences between obtained and expected frequencies within the cells are considered to be a reflection of a genuine difference between the categories of the variable. This difference indicates that a statistically significant relationship exists between the variables (that is, political party does make a difference in gun control opinion in the general population), not just the sample. If the calculated chi

EXHIBIT 10.1. CRITICAL VALUES OF THE CHI-SQUARE DISTRIBUTION.

df	$\chi^{2*}_{.05}$	$\chi^{2*}_{.01}$
1	3.841	6.635
2	5.991	9.210
3	7.815	11.345
4	9.488	13.277
5	11.070	15.086
6	12.592	16.812
7	14.067	18.475
8	15.507	20.090
9	16.919	21.666
10	18.307	23.209
11	19.675	24.725
12	21.026	26.217
13	22.362	27.688
14	23.685	29.141
15	24.996	30.578
16	26.296	32.000
17	27.587	33.409
18	28.869	34.805
19	30.144	36.191
20	31.410	37.566
21	32.671	38.932
22	33.924	40.289
23	35.172	41.638
24	36.415	42.980
25	37.652	44.314
26	38.885	45.642
27	40.113	46.963
28	41.337	48.278
29	42.557	49.588
30	43.773	50.892

square is less than the critical chi square, then no relationship between the variables has been identified, which in this case would allow the researcher no immediate alternative other than to operate on the basis that no genuine relationship exists between political party and gun control opinion.

The process of identifying the critical chi square for the contingency table (Table 10.4) requires two pieces of information. First, the researcher must decide the appropriate level of confidence (generally 95 percent or 99 percent). Second, degrees of freedom must be determined. The simple rule for degrees of freedom for chi square involves the following formula:

$$df = (r - 1)(c - 1)$$

(10.2)

where

df = degrees of freedom

r = number of categories of the dependent variable (row variable)

c = number of categories of the independent variable (column variable)

Degrees of freedom can be explained in the context of chi square as the number of cells that are free to vary. Once the values of this number of cells are known and all row and column totals are known, the values of all other cells become fixed. In the context of Table 10.4, equation 10.2 indicates that $df = 4$. Thus, with the row and column totals identified and the frequencies of any four cells known, the remaining cell frequencies can be calculated.

The critical chi square for Table 10.4, at the 95 percent level of confidence and with $df = 4$, can be determined from Exhibit 10.1 to be 9.488. For the 99 percent level of confidence, the critical chi square is 13.277. Therefore, the calculated chi square of 2.23 is less than the critical chi square and is indicative of sampling error rather than of a statistically significant relationship between the variables. Because the researcher cannot establish a statistical relationship, he or she will proceed under the assumption that there is none.

In another example of the use of chi square, consider Table 10.2, where it appears that there is a difference among various tribes in their time living on the reservation. The calculated chi square for this table is 23.712, and at 10 df, this is sufficiently large (greater than the critical chi squares of 18.307 and 23.209) to indicate a statistically significant relationship between the variables at the 95 percent and 99 percent levels of confidence, respectively. In other words, length of residence on reservations has been found to be related to tribe, and the apparent longer residency of the San Pasqual tribe in the sample is generalizable to the full populations on these reservations.

Cramér's *V* and Phi (ϕ)

Tests of statistical significance determine whether a relationship exists between variables, but they do not measure the strength of that relationship. Measures of association reflect the strength of the relationship between two or more variables. They are very useful as descriptive tools to indicate the strength of a relationship and as qualifying devices for determining whether or not certain findings merit reporting.

In the case of a very large survey sample with margins of error, for example, of ±1 percent or less, there likely will be many findings of statistical significance. A study we once performed contained 55,000 surveys—or a margin of error of ±0.4 percent. Differences as small as 1 percent therefore were found to be statistically significant. To have reported all such findings of significance would have made the final report unreadable. Using measures of association as a qualifying device allowed the report to include only findings that had achieved a certain strength.

There are measures of association that can be derived for nominal data directly from the calculated chi-square statistic. The most versatile of these measures is Cramér's *V*, for which the formula is as follows:

$$V = \sqrt{\frac{\chi^2}{n(M-1)}} \tag{10.3}$$

where

χ^2 = calculated chi square

n = sample size

M = minimum number of rows or columns

The possible values for Cramér's *V* range from 0 to 1, with 0 representing no association and 1 representing a perfect association.

Table 10.5 depicts the results of a survey of 550 respondents to a questionnaire about the value of psychiatric care. The table seems to show that

TABLE 10.5.   VALUE OF PSYCHIATRIC CARE, BY AGE OF RESPONDENT.

	Age of Respondent							
	Under 35		35–55		Over 55		Total	
Value	f	%	f	%	f	%	f	%
Always helpful	15	7.8	21	10.1	8	5.3	44	8.0
Sometimes helpful	68	35.4	135	64.9	85	56.7	288	52.4
Rarely helpful	98	51.0	50	24.0	46	30.7	194	35.3
Never helpful	11	5.8	2	1.0	11	7.3	24	4.3
Total	192	100.0	208	100.0	150	100.0	550	100.0

middle-aged people are more favorably inclined toward psychiatry than are the younger and the older groups.

A chi-square test found a significant calculated chi square of 51.02, indicating that middle-aged people are in fact more favorably inclined toward psychiatric care. Cramér's V can add further information by telling the researcher how strong the relationship between age and opinion about psychiatry is. Cramér's V is calculated in the following manner:

$$V = \sqrt{\frac{51.02}{550(3-1)}}$$

$$= \sqrt{\frac{51.02}{1,100}}$$

$$= \sqrt{.0464}$$

$$= .22$$

Exhibit 10.2 represents a scale for interpreting the meaning of Cramér's V. From Exhibit 10.2, it can be found that the statistically significant relationship between age and opinion concerning psychiatric care can be labeled as moderately strong. Note that most significant relationships are found to be moderate or relatively strong at best and that Cramér's V rarely achieves a value of .80 or above.

A special case of Cramér's V is phi (ϕ). Phi is the measure of association based on the chi-square distribution when one or both of the variables contains only two categories. The formula for phi is as follows:

$$\phi = \sqrt{\frac{\chi^2}{n}} \tag{10.4}$$

Note that the formula for phi is the same as the formula for Cramér's V, with at least one variable containing only two categories. Therefore, phi is also interpreted in accordance with the scale represented by Exhibit 10.2.

EXHIBIT 10.2. INTERPRETATION OF CALCULATED CRAMÉR'S V, PHI, AND LAMBDA MEASURES OF ASSOCIATION.

Measure	Interpretation
.00 and under .10	Negligible association
.10 and under .20	Weak association
.20 and under .40	Moderate association
.40 and under .60	Relatively strong association
.60 and under .80	Strong association
.80 to 1.00	Very strong association

Additional Chi-Square Considerations

The use of chi square is subject to certain restrictions. As is true of all other sampling techniques and the results derived from them, the chi-square statistic is more reliable as the overall sample size increases. Consistent with this principle is the rule of thumb that each cell of the contingency table should contain an expected frequency of at least 5. If the expected frequency falls below 5 in any one cell, categories should be merged to eliminate the problem. This should be done in accordance with logic and reasonableness.

For example, a survey question that asked people to identify their dominant mode of travel contained the following categories: automobile, large bus, rail, taxi, minibus, dial-a-ride, and car pool. Let us suppose that the cells associated with the minibus and dial-a-ride categories contained expected frequencies of less than 5. It could be appropriate to combine these categories into a single category with an expected frequency of 5 or more because each of these forms of travel carries passengers on street-oriented systems in relatively small, public vehicles. Alternatively, a researcher might decide to combine the minibus category with the bus category, since both have similar mass transit features, and combine the taxi and dial-a-ride categories, since each of these modes of travel is associated with placing a telephone call for service. In general, however, the researcher's goal is to maintain as many of the original question categories as possible; therefore, the process of combining categories should be undertaken only when necessary to achieve requisite expected frequencies.

Contingency tables consisting of two categories of independent variables and two categories of dependent variables (commonly referred to as 2×2 tables) present the researcher with additional concerns. For example, if a 2×2 table contains cells with expected frequencies of less than 5, no merging of categories is possible. The researcher must employ a somewhat different test of statistical differences, known as *Fisher's exact test*. Also, results in a 2×2 table can be distorted if any one cell contains an expected frequency of less than 10. In such cases, it is recommended that the *Yates correction* be employed. This simply entails reducing the magnitude of the difference between observed frequencies and expected frequencies in each of the four cells by 0.5. Most computer software programs perform this correction automatically.

When either variable is on the ordinal or interval scale, the use of chi square can become problematic as the researcher attempts to interpret the results. Consider Table 10.6, which indicates overall political orientation of local businesspersons on an ordinal continuum from "very liberal" to "very conservative," by gender.

The calculated chi square for Table 10.6 is statistically significant ($\chi^2 = 18.40$), indicating a relationship between political orientation and gender. An initial glance

TABLE 10.6. POLITICAL ORIENTATION OF LOCAL BUSINESSPERSONS, BY GENDER.

Political Orientation	Male		Female		Total	
	f	%	f	%	f	%
Very liberal	20	20.0	20	20.0	40	20.0
Liberal	25	25.0	10	10.0	30	15.0
Moderate	0	0.0	10	10.0	15	7.5
Conservative	30	30.0	40	40.0	70	35.0
Very conservative	25	25.0	20	20.0	45	22.5
Total	100	100.0	100	100.0	200	100.0

at the table might lead the researcher to think that men are more conservative than women because 25 percent of the men consider themselves very conservative, whereas only 20 percent of the women do so. However, further review of the table shows 60 percent of the women and only 55 percent of the men in both conservative categories combined. The review also shows 45 percent of the men in the two liberal categories, while only 30 percent of the women are so categorized. Thus, it is not clear, from an initial review of the table, whether there is a consistent pattern identifying a particular gender with either a liberal or a conservative orientation.

A significant chi square can result from data that may possess major differences between obtained and expected frequencies in as few as two cells. In this example, these differences are most pronounced in the "moderate" cells. When using nominal data, valid conclusions can still be drawn from such differences, but when the data are based on a continuum (as is the case in Table 10.6), the objective is to detect an overall pattern. In the case of Table 10.6, there is no such clearly identifiable pattern. A significant chi square that results from two or more major discrepancies within the cells of ordinal or interval data can therefore lead the researcher to misinterpret the overall meaning of the data. Hence, additional tests of significance may be necessary to evaluate significant relationships among ordinal and interval variables.

Gamma (γ)

The nature of chi square, a test built for nominal data, is such that significance is achieved if cells differ sufficiently from their null hypothesis values regardless of logical order. As such, when trends are being analyzed (for example, a hypothesis that as incomes increase, people become more conservative politically), chi square may not be the best test of significance available. Gamma generally proves to be a better test. Gamma is not a significance test; rather, it is a measure of association (as is Cramér's V) that has a Z significance test that can be performed along with it.

When each of two variables is on the ordinal or interval scale, gamma (γ) is an appropriate measure of association and one that measures the strength of trends in the data. The formula for gamma is

$$\gamma = \frac{\Sigma (f_i \cdot \Sigma f_s) - \Sigma (f_i \cdot \Sigma f_d)}{\Sigma (f_i \cdot \Sigma f_s) + \Sigma (f_i \cdot \Sigma f_d)} \tag{10.5}$$

where

f_i = the frequency of any cell

f_s = the frequency of a cell ordered in the same direction from the subject cell

f_d = the frequency of a cell ordered in a different (or inverse) direction from the subject cell

Consider Table 10.7, which results from cross-tabulating two survey questions: one eliciting information about the educational level of the college graduate respondents to the survey and the other about the respondent's socioeconomic status.

Operationalizing the equation for gamma requires ensuring that both variables are ordered similarly. That is, both variables should be presented in order, from high to low or from low to high. Having verified that each variable follows the same pattern, the researcher can determine $\Sigma (f_i \cdot \Sigma f_s)$ by taking the frequency in each cell and multiplying that number by the frequencies in each category cell that is both below that cell and to the right of it—in other words, cells for which both variables are either lower in rank or higher in rank than the cell under consideration. In the case of Table 10.7, $\Sigma (f_i \cdot \Sigma f_s)$ can

TABLE 10.7.   EDUCATIONAL LEVEL BY SOCIOECONOMIC STATUS.

| Socioeconomic Status | Educational Level | | | | | | | |
| | Doctorate | | Master's | | Bachelor's | | Total | |
	f	%	f	%	f	%	f	%
Upper	25	62.5	20	20.0	5	8.3	50	25.0
Middle	10	25.0	55	55.0	30	50.0	95	47.5
Lower	5	12.5	25	25.0	25	41.7	55	27.5
Total	40	100.0	100	100.0	60	100.0	200	100.0

be calculated as follows:

$$\Sigma(f_i \cdot \Sigma f_s) = 25(55 + 25 + 30 + 25) = 3,375$$
$$+ 10(25 + 25) \quad = \quad 500$$
$$+ 20(30 + 25) \quad = 1,100$$
$$+ 35(25) \quad = 1,375$$
$$\overline{\qquad 6,350}$$

The determination of $\Sigma(f_i \cdot \Sigma f_d)$ requires taking the frequency in each cell and multiplying that number by the frequencies in each category cell that is both below and to the left of it—below and left being inverse directions, with one variable higher in rank and the other variable lower. Again from Table 10.7:

$$\Sigma(f_i \cdot \Sigma f_d) = 5(55 + 25 + 10 + 5) = 475$$
$$+ 30(5 + 25) \quad = 900$$
$$+ 20(10 + 5) \quad = 300$$
$$+ 55(5) \quad = 275$$
$$\overline{\qquad 1,950}$$

Hence, gamma can be calculated as follows:

$$\gamma = \frac{6,350 - 1,950}{6,350 + 1,950} = \frac{4,400}{8,300} = +.53$$

The calculated gamma of +.53 is interpreted on a scale that ranges from −1.00 to +1.00. Measures in the positive range are indicative of variables that vary in the same direction; that is, as one increases (decreases), the other increases (decreases). In Table 10.7, for instance, the higher the level of education is, the higher is the socioeconomic status. Measures in the negative range imply variables that vary in an inverse manner: while one increases, the other decreases. The scale presented in Exhibit 10.3 can serve as a guideline for interpreting the calculated gamma measure, which in the case of Table 10.7 can be thought of as moderately positive in nature.

Testing the Statistical Significance of Gamma

Any apparent statistical pattern or trend that the researcher observes in the data must be formally tested for statistical significance. The chi square serves as the test for the significance of Cramér's *V*, phi, and lambda, but the presence of ordinal

EXHIBIT 10.3. INTERPRETATION OF CALCULATED GAMMA.

Measure	Interpretation
−1.00	Perfect inverse association
−.75 to −.99	Very strong inverse association
−.60 to −.74	Strong inverse association
−.30 to −.59	Moderate inverse association
−.10 to −.29	Low inverse association
−.01 to −.09	Negligible inverse association
0	No association
+.01 to +.09	Negligible positive association
+.10 to +.29	Low positive association
+.30 to +.59	Moderate positive association
+.60 to +.74	Strong positive association
+.75 to +.99	Very strong positive association
+1.00	Perfect positive association

or interval data used in the calculation of gamma permits the use of a more direct test of significance for the calculated gamma.

The following equation is used to calculate the Z score necessary for determining the significance of gamma:

$$Z = \gamma \left(\sqrt{\frac{|\Sigma(f_i \cdot \Sigma f_s) - \Sigma(f_i \cdot \Sigma f_d)|}{n(1 - \gamma^2)}} \right) \tag{10.6}$$

Applying equation 10.6, the gamma derived from Table 10.7 yields

$$Z = .53 \left(\sqrt{\frac{4,400}{200(1 - (.53)^2)}} \right)$$

$$= .53 \left(\sqrt{\frac{4,400}{143.82}} \right)$$

$$= 2.93$$

Inasmuch as gamma can be either positive or negative, the significance test for gamma's Z score is a two-tail test, with critical Z scores of 1.96 (95 percent confidence) and 2.575 (99 percent confidence). As such, the calculated Z of 2.93 is indicative of significance at both the 95 percent and 99 percent levels of confidence.

It is noteworthy that the significance test for gamma can be used to test the significance of ordinal data in contingency tables when the results of the chi-square test are not definitive because of the insensitivity of chi square to patterns and trends, as discussed previously.

EXERCISES

1. The following matrix of cross-tabulated response frequencies has been provided in a computer printout. The categories are ethnicity and political party preference:

	White	Black	Hispanic	Asian	Total
Democrat	200	120	100	50	470
Republican	350	10	40	30	430
Independent	50	20	10	20	100
Total	600	150	150	100	1,000

   a. Use the data from this matrix to construct a contingency table. Make certain that the independent and dependent variables are appropriately placed in the columns and rows.
   b. What conclusions might you draw from this table?
   c. Test the statistical significance of your conclusions.

2. A random sample of 500 registered voters was conducted to find out the relationship between region of the country and whether respondents felt the United States should have approved the NAFTA treaty (concerning free trade with Mexico and Canada). The findings were as follows:
   - Among those in favor, 70 were from states west of or including the Rocky Mountains, 53 were from the central part of the country, 100 were from the East, and 50 were from the South.
   - Among those opposed, 35 were from the West, 67 were from the central section, 75 were from the East, and 50 were from the South.

   Prepare a contingency table detailing these findings.

3. A random sample of 400 registered voters was conducted to find out the relationship between religion and whether the United States should enter into serious negotiations with the Palestine Liberation Organization (PLO) concerning a homeland for Palestinians. The findings are reported below:
   a. What preliminary conclusions might you draw from Table 10.8?
   b. Determine whether your preliminary conclusions are statistically significant.

4. The results from a survey of 350 respondents are depicted in Table 10.9.
   a. Indicate the strength and direction of the relationship between number of children and level of education of the parent.
   b. Is this relationship statistically significant?

5. Use the data from Exercise 3 to indicate the strength of the association between religion and opinion regarding PLO negotiations.

TABLE 10.8. OPINION CONCERNING NEGOTIATIONS WITH THE PALESTINE LIBERATION ORGANIZATION, BY RELIGIOUS PREFERENCE.

	Jewish		Catholic		Protestant		Other		Total	
Opinion	f	%	f	%	f	%	f	%	f	%
Yes	10	20.0	30	30.0	120	60.0	40	80.0	200	50.0
No	40	80.0	70	70.0	80	40.0	10	20.0	200	50.0
Total	50	100.0	100	100.0	200	100.0	50	100.0	400	100.0

TABLE 10.9. NUMBER OF CHILDREN BY LEVEL OF EDUCATION OF PARENT.

	Level of Education							
	Postgraduate Degree		Bachelor's Degree		High School Diploma or Less		Total	
Number of Children	f	%	f	%	f	%	f	%
0	12	40.0	30	25.0	30	15.0	72	20.6
1	8	26.7	30	25.0	40	20.0	78	22.3
2	6	20.0	30	25.0	50	25.0	86	24.6
3+	4	13.3	30	25.0	80	40.0	114	32.5
	30	100.0	120	100.0	200	100.0	350	100.0

6. Use the data from Exercise 4 to indicate the strength of association between income and attitude toward growth management.

7. Table 10.10 depicts the results of a sample survey in which people were asked, among other issues, to indicate their level of support for an urban light rail transit system to a major local university. These opinions were cross-tabulated by the level of education of the respondent.

   a. Is there a statistically significant relationship between education level and opinion concerning light rail?

   b. If so, describe the strength (and direction, if applicable) of that relationship using the most appropriate measure of association.

8. The data in Table 10.11 have been derived from a study you recently undertook.

   a. Write a brief few sentences telling someone who looks at this table what he or she should understand from it.

   b. Is there a statistically significant relationship between recreational choice and state of residence? Perform the appropriate test of significance.

TABLE 10.10. SUPPORT FOR URBAN LIGHT RAIL
TO UNIVERSITY BY LEVEL OF EDUCATION.

	Education					
	Graduate Degree		College Graduate		High School or Less	
Support	f	%	f	%	f	%
Disapprove	18	18.0	43	23.4	97	43.3
Approve somewhat	22	22.0	67	36.4	68	30.4
Approve strongly	60	60.0	74	40.2	59	26.3
Total	100	100.0	184	100.0	224	100.0

TABLE 10.11. FAVORITE RECREATIONAL ACTIVITY
BY STATE OF RESIDENCE.

	State					
	New York		California		Texas	
Activity	f	%	f	%	f	%
Snowboarding	75	41.7	105	47.7	15	15.0
Running[a]	55	30.6	95	43.2	30	30.0
Bowling	50	27.7	20	9.1	55	55.0
Total	180	100.0	220	100.0	100	100.0

[a]Either for recreation or to escape physical assault

    c.   Which variable is your independent variable, according to the Table 10.11?

    d.   Describe the strength of that relationship, if a relationship exists at all.

9.   A random sample of 1,100 City of San Diego voters was interviewed by telephone to find out how they feel about a proposal in San Diego to sell the naming rights for the area of downtown around the new downtown ballpark to corporate sponsors. No city in the United States has ever sold naming rights to a portion of its city. Among 110 people in the downtown area, 40 thought it was a good idea, and 70 thought that it was not. Among 330 residents who lived south of I-8, 220 liked the idea and 110 did not. There were 440 residents of the western portion of the city who were split 220 to 220, and voters along I-15 were somewhat supportive, by a vote of 120 to 100.

    a.   Prepare a contingency table detailing these findings.

    b.   Examine your table, and explain what you believe it says about the support for the proposal and its geographical constituencies.

TABLE 10.12. VOTER PREFERENCE REGARDING GROWTH MANAGEMENT PLAN FOR THE CITY OF WILLOWBEND BY ANNUAL HOUSEHOLD INCOME.

Attitude Toward Growth Management Plan	Annual Household Income												
	Under $15,000		$15,000 and under $30,000		$30,000 and under $60,000		$60,000 and under $100,000		$100,000 and Above		Total	
	f	%	f	%	f	%	f	%	f	%	f	%
Favor	8	66.7	81	88.0	48	25.3	13	18.3	9	28.0	159	39.8
Do not favor	1	8.3	9	9.8	130	68.4	50	70.4	18	48.0	208	52.0
No opinion	3	25.0	2	2.2	12	6.3	8	11.3	8	24.0	33	8.2
Total	12	100.0	92	100.0	190	100.0	71	100.0	35	100.0	400	100.0

    c.   Is your hypothetical interpretation from *b* generalizable to the total population of San Diego voters, or is it merely the routine difference between a sample and its population (sampling error)?

    d.   How would you characterize the strength of the relationship, if any?

10.   The city of Willowbend conducted a sample survey of 400 registered voters to determine whether voters have a favorable attitude toward the city's newly proposed growth management plan. The city was interested not only in voter preference but also in how this preference relates to voter income level. Accordingly, the sample respondents were asked to indicate their annual household income among certain fixed categories. The results of the income question and the growth management preference question were cross-tabulated. The results of that cross-tabulation are presented in Table 10.12.

        Use the chi-square test to determine whether there is a statistically significant relationship (95 percent level of confidence) between voter attitude concerning the growth management plan and annual household income. *Hint:* Consider sparse cells, and regroup categories accordingly before calculating the chi square.

Notes

1. (This endnote is best understood after reading the entirety of Chapter Ten.) Another measure of association associated with nominal scale data is lambda (λ). The formula for lambda is as follows:

$$\lambda = \frac{\sum F_{iv} - R_{dv}}{n - R_{dv}} \tag{10.7}$$

where

    F_{iv} = largest cell frequency for each category of the independent variable
    R_{dv} = largest row total among the categories of the dependent variable
    n = sample size

    Table 10.13 depicts the results of a survey of 510 residents of a metropolitan area in Rhode Island. Respondents were asked to indicate their most frequent method of commuting to work. Results of this question were cross-tabulated by subarea.

    Applying equation 10.7 to these data yields the calculation for lambda, as follows:

$$\sum F_{iv} = 68 + 70 + 103 + 100 = 341$$
$$R_{dv} = 273$$
$$n = 510$$
$$\lambda = \frac{341 - 273}{510 - 273}$$
$$= \frac{68}{237}$$
$$= .29$$

TABLE 10.13.   MOST FREQUENTLY USED
MODE OF TRANSPORTATION TO WORK BY SUBAREA.

	Inner City		Mid-City		Suburban		Exurban		Total	
	f	%	f	%	f	%	f	%	f	%
Auto	20	17.0	50	31.3	103	83.1	100	92.6	273	53.5
Bus	68	57.6	70	43.7	16	12.9	4	3.7	158	31.0
Bicycle/walk	30	25.4	40	25.0	5	4.0	4	3.7	79	15.5
	118	100.0	160	100.0	124	100.0	108	100.0	510	100.0

WORKSHEET 10.2.   LAMBDA AS A MEASURE
OF REDUCTION IN PREDICTIVE ERROR.

Subarea	Predicted Transportation Method (Modal Category for Subarea)	Number of Predictive Errors
Inner City	Bus	(20 + 30) out of 118
Mid-City	Bus	(50 + 40) out of 160
Suburban	Auto	(16 + 5) out of 124
Exurban	Auto	(4 + 4) out of 108
Total		169 out of 510

Chance of error knowing independent variable = 169/510 = 33.1 percent.

Chance of error not knowing independent variable (from discussion above) = 46.5 percent.

Reduction in error = 46.5 percent – 33.1 percent = 13.4 percent.

Proportionate reduction in error = 13.4 percent/46.5 percent = .29 = λ.

The calculated lambda can also be interpreted according to Exhibit 10.2 as indicative of a moderate association between metropolitan subareas and mode of travel to work. As with Cramér's V and phi, lambda values can range from 0 (no association) to 1 (perfect association).

Beyond its usefulness as a measure of association, lambda also measures the extent to which the independent variable serves to explain the variation in the dependent variable. In the example in Table 10.13, a calculated lambda of .29 indicates that the error associated with predicting values of the dependent variable is reduced by 29 percent when the value of the independent variable is known.

For example, for the data in Table 10.13, if a researcher had only the knowledge of transportation mode for the entire metropolitan area (not disaggregated by subarea), he or she would tend to predict that any one individual is most likely to use an automobile to travel to work, because "auto" is the modal category. This prediction, however, has only a 53.5 percent chance (273/510) of being correct and, conversely, a 46.5 percent chance ([158 1 79]/510) of being incorrect. However, the researcher's chance of being incorrect decreases with the knowledge of subarea modes of transportation. This decrease can be illustrated by Worksheet 10.2.

Worksheet 10.2 illustrates that knowledge of the values of the independent variable has reduced the chance of predictive error from 46.5 percent to 33.1 percent—a 13.4 percent reduction in error. In relation to the original error of 46.5 percent, this represents a proportionate reduction of .29 (13.4 percent/46.5 percent)—precisely the value of lambda. Hence, not only is lambda a measure of association between two nominal scale variables, it also measures the extent to which the independent variable explains the dependent variable.

This explanatory power of lambda is particularly useful in cases where the researcher has not yet identified, for whatever reason, which variable is the independent variable and which is the dependent variable. The determination of which variable provides the better reduction in error, and is hence the more likely independent variable, can be readily determined applying equation 10.7 twice, with each variable alternating as the independent variable. The greater of the two calculated lambdas will indicate the more appropriate independent variable. The reader can verify that subarea is the more appropriate independent variable for the data in Table 10.13 by calculating the alternative lambda to equal .15.

Lambda is less adequate than Cramér's V in the fact that lambda measures association in only one direction—how the independent variable affects the dependent variable. Cramér's V measures the total relationship between the two variables, and relationships are hardly ever one-way. In the metropolitan area/transportation method example, it is likely that the independent variable (subarea) does have a greater impact on transportation than the reverse; however, there can be no doubt that this is not exclusively one-way and that some individuals will choose a place to live because of transportation considerations. Cramér's V will capture both relationships, whereas lambda will measure only the impact of area on transportation choice.

2. The frequency percentages should always be shown to sum to 100 percent. If, due to rounding error, the percentages total slightly less or slightly more than 100 percent, the researcher can address the problem in one of two ways. First, she or he may show a total percentage of 100 percent even if the individual percentages result in a slightly different sum; a footnote should be added to the table stating that percentages may not add to 100 percent because of rounding error.

Second, a category can be identified with a percentage that can be adjusted to achieve a total of 100 percent. It is recommended that the category chosen be the one that is associated with the percentage closest to the point at which rounding takes place. For example, in Table 10.1, the category "Less than 1 year" is associated with a percentage of exactly 13.049. Properly rounded, this percentage should read 13.0. In so doing, however, the total percentage for the entire adjusted frequency distribution equals 99.9 percent. The closeness of 13.049 to the rounding point of 13.05 warrants its selection as the category to be adjusted.

We recommend the second approach, because the slight advantage in accuracy of using 13.0 percent is far less important than the disadvantage of the appearance of inaccuracy resulting from a column of percentages that does not add up to the total percentage indicated. The primary purpose of tabular construction is to communicate data visually with simplicity and clarity, and the second approach more readily achieves this objective.

CHAPTER ELEVEN

TESTING THE DIFFERENCE BETWEEN MEANS

Chapter Ten discussed analytical statistics for data that lend themselves to presentation in the form of contingency tables. Contingency table data typically are not interval in level of measurement because the very large number of individual values associated with interval data would render the table unwieldy and unreadable. This chapter continues the discussion of significance tests that are most appropriately applied when the dependent variable is interval, on ordinal scale treated as interval (see Chapter Seven), or on ordinal containing categories of grouped interval data (Chapters Three and Five).

These significance tests are more powerful statistical tools, with wider ranging applicability than the tests discussed in the previous chapter.

Independent Samples *t* Test

When the dependent variable is on the interval scale and an independent variable consists of only two categories on any scale, there is a test of statistical significance that is more rigorous than the chi-square test of significance or Gamma's Z. This test involves the comparison of the arithmetic means of the two categories of the independent variable, and it is known as the *independent samples t test*.

Table 11.1 reflects two distinct sample groups of government employees who sat for a promotional examination. A review of these data would lead the

TABLE 11.1. PROMOTIONAL EXAMINATION SCORE BY EMPLOYER (CITY OR COUNTRY).

Test Scores	City Employees		County Employees		Total	
	f	%	f	%	f	%
90–100	30	15.0	20	10.0	50	12.5
80–89	65	32.5	35	17.5	100	25.0
70–79	60	30.0	65	32.5	125	31.3
60–69	30	15.0	30	15.0	60	15.0
50–59	10	5.0	30	15.0	40	10.0
40–49	5	2.5	20	10.0	25	6.2
	200	100.0	200	100.0	400	100.0
	$\bar{x}_1 = 77.6$		$\bar{x}_2 = 70.8$			
	$s_1 = 12$		$s_2 = 14.5$			

researcher to suspect that county employees tend to perform better than city employees (note the one-tail question; see Chapter Seven), with the mean score for county employees almost seven points higher than for city employees. This suspicion should prompt the researcher to test whether this apparent difference is genuine or the result of sampling error.

In the case at hand, the dependent variable, being on the interval scale, will permit the researcher to employ the independent samples t test. The equation for the test is as follows:

$$t = \frac{\bar{x}_1 - \bar{x}_2}{\sqrt{\dfrac{s_1^2}{n_1 - 1} + \dfrac{s_2^2}{n_2 - 1}}} \tag{11.1}$$

Application of the equation to the data in Table 11.1 yields

$$t = \frac{77.6 - 70.8}{\sqrt{\dfrac{(12)^2}{199} + \dfrac{(14.5)^2}{199}}}$$

$$t = 5.1$$

Degrees of freedom are $n - 1$ for each group, so that the total degrees of freedom are

$$df = n_1 + n_2 - 2, \tag{11.2}$$

which, when applied to this case, equals 398.

Exhibit 7.2 can be consulted, and it is seen that $t = 5.1$ exceeds the critical t for one-tail questions at both the 95 percent (critical $t = 1.645$) and 99 percent (critical $t = 2.33$) confidence levels. Note that when df exceeds 120, the critical t and critical Z (see Chapter Seven) are considered to effectively coincide. The independent samples t test can be used to test the difference between any two arithmetic means, whether or not these means derive from a table, such as Table 11.1. For example, suppose that 6,000 individuals across the United States are asked how many miles they commute to work. The 2,000 respondents from west of the Mississippi River indicate a mean of 6.25 miles (standard deviation $= 2.12$ miles). The 4,000 respondents from east of the Mississippi reveal a mean commute of 6.03 miles (standard deviation $= 1.3$ miles). Can the researcher conclude that a significant difference (two-tail question) exists based on geographic considerations?

Applying equation 11.1 provides the following calculations:

$$t = \frac{6.25 - 6.03}{\sqrt{\dfrac{(2.12)^2}{1,999} + \dfrac{(1.31)^2}{3,999}}} = \frac{0.22}{\sqrt{0.0022 + 0.0004}} = \frac{0.22}{0.05} = 4.4$$

A calculated t of 4.4 exceeds both the 95 percent confidence level critical two-tail t of 1.96 and the 99 percent confidence level critical t of 2.575. Hence, the researcher can conclude with 99 percent confidence that the two sections of the country differ significantly in their commuting mileage.

Paired Samples t Test

The independent samples t test assumes that the variances for each sample are different. That is, they are independent samples. There are situations when two samples are not independent. A before-and-after test is a situation where the second sample (after) is dependent on the first (before). For example, if an exercise program were to have been introduced for the Community B fire department depicted in Table 5.6 (Chapter Five) and the 6.0 minute mean endurance time were increased to 7.0 minutes for a sample of 25 firefighters with a standard deviation of 3.0 (recall that Table 5.6 indicates $n = 200$ and $s = 2.82$), the fire chief would be interested to determine if this increase were statistically significant (one-tail). These 25 firefighters were chosen at random from among the 200 who took the test originally and for whom individual performance data from that first test are no longer available. Equations 11.3 and 11.4 would provide this information.

First, calculate the overall standard deviation of the paired samples:

$$s = \sqrt{((n_1 - 1)s_1^2 + (n_2 - 1)s_2^2) \div (n_1 + n_2 - 2)} \qquad (11.3)$$

Then calculate the t test, as follows:

$$t = (\bar{x}_1 - \bar{x}_2) \div s\sqrt{(1/n_1 + 1/n_2)} \qquad (11.4)$$

In this case, the calculation would yield

$$s = \sqrt{((24(9) + 199(7.952)) \div (25 + 200 - 2))} = 2.84$$
$$t = (7.0 - 6.0) \div 2.84\left(\sqrt{1/25 + 1/200}\right)$$
$$= 1.0 \div 2.84\left(\sqrt{.045}\right)$$
$$= 1.0 \div .602$$
$$= 1.66$$

The one-tail (looking for significant increase) critical t (from Exhibit 7.2, $-df = n_1 + n_2 - 2$) is 1.645 at 95 percent confidence; therefore, the exercise program did improve performance.

Difference of Proportions Test

There exists a variation on the independent samples t test that is applicable to percentage values for the dependent variable for any two categories of the independent variable. This test is very specific to only two independent variable categories. Therefore, for the vast majority of analyses of survey data that are nominal or ordinal, chi-square and Gamma's Z are still the significance tests of choice.

Frequently survey data from two groups are analyzed in conjunction with proportions and percentages rather than arithmetic means. This is particularly so in cases where ordinal and nominal data are present—data that do not lend themselves to the calculation of means. For instance, a sample survey of 267 respondents in Kalamazoo, Michigan, finds that 45 percent of the respondents have trust in their public leaders. A similar survey of 320 respondents in Lansing, Michigan, finds that 38 percent have such trust. A researcher decides to evaluate this apparent difference and seeks initially to ascertain statistical significance. The researcher poses this question: Is Kalamazoo more trusting of its public leaders than is Lansing?

The response to this research question can be derived by application of the difference of proportions test:

$$t = \frac{\overline{p}_1 - \overline{p}_2}{\sqrt{\left(\dfrac{(n_1 - 1)\overline{p}_1 + (n_2 - 1)\overline{p}_2}{n_1 + n_2 - 2}\right)\left(1 - \dfrac{(n_1 - 1)\overline{p}_1 + (n_2 - 1)\overline{p}_2}{n_1 + n_2 - 2}\right)\left(\dfrac{n_1 + n_2 - 2}{(n_1 - 1)(n_2 - 1)}\right)}} \tag{11.5}$$

where

$\overline{p}_1$ = sample proportion of first subgroup

$\overline{p}_2$ = sample proportion of second subgroup

n_1 = sample size of first subgroup

n_2 = sample size of second subgroup

The calculated t is then compared to the same critical t values as in the independent samples t test to establish the presence or absence of statistical significance. This process can be illustrated as follows:

$$t = \frac{0.45 - 0.38}{\sqrt{\left(\dfrac{266(0.45) + 319(0.38)}{266 + 319}\right)\left(1 - \left(\dfrac{266(0.45) + 319(0.38)}{266 + 319}\right)\right)\left(\dfrac{(266 + 319)}{(266)(319)}\right)}}$$

$$= \frac{0.07}{\sqrt{\left(\dfrac{119.7 + 121.22}{585}\right)\left(1 - \left(\dfrac{119.7 + 121.22}{585}\right)\right)\left(\dfrac{585}{84,854}\right)}}$$

$$= \frac{0.07}{\sqrt{(0.412)(0.588)(0.0069)}}$$

$$= \frac{0.07}{0.041}$$

$$= 1.71$$

Because the calculated t of 1.71 exceeds the one-tail, 95 percent critical t of 1.645, the researcher can conclude, with 95 percent confidence, that Kalamazoo residents trust their public leaders more than the residents of Lansing trust theirs.

Analysis of Variance

These independent and paired samples *t* tests compare two categories (subgroups) of the independent variable—for instance, men versus women or a before group to the same individuals after some change has been introduced. The world, however, is not always so easily categorized into only two groups, and even when it is, it is often in the researcher's interest to analyze the data in greater detail. In the case of ethnicity, the researcher's purposes might best be served by replacing two categories, such as "white" and "nonwhite," with "white," "African American," "Latino," and "Asian." When there are multiple categories of the independent variable, with a dependent variable on the interval scale, the researcher can attempt to determine statistical significance by a series of applications of these means tests or by a single application of an analysis of variance test.

Besides the cumbersome nature of a series of the means tests, there is an even more important shortcoming to that option. Repeated applications of a test to the same data increases the likelihood of committing a Type I error, with each test adding its 5 percent error factor (at 95 percent confidence) to previous applications. Hence, there is a significant advantage to identifying and implementing one single test with its own 5 percent Type I error factor only. That single test is the *analysis of variance test.*

The analysis of variance test measures the amount of the total variability of the dependent variable that can be attributed to the differences among the categories of the independent variable. Analysis of variance compares that portion of the variance in the data that can be attributed to the independent variable categories to that portion of the variance that is attributable to all other potential factors. For example, suppose the researcher is interested in determining if a significant difference exists among ethnic groups with regard to income. Analysis of variance would measure the amount of all the variation in the data that can be explained by ethnic differences as opposed to such other factors as education, marital status, and age. The relative importance of this one independent variable is determined by the analysis of variance test and is tested for statistical significance.

The implementation of the analysis of variance test requires dividing the total variation associated with the dependent variable into two parts: (1) between-group variation (that portion of the variation explained by grouping the dependent variable in a particular fashion) and (2) within-group variation (that portion of the variation accounted for by all factors other than the groups). The test is not dissimilar from the means tests that measure distance between group means and divides this distance by a factor that is indicative of the variation within the groups themselves.

Consider the following example of hourly wage rates paid to a sample of public works employees in four cities: Harrisburg, Utica, Annapolis, and Roanoke:

Harrisburg	Utica	Annapolis	Roanoke
$12.60	$13.80	$15.00	$12.60
$10.40	$15.60	$16.00	$10.60
$16.00	$10.40	$15.60	$12.60
$12.60	$10.40	$15.40	$10.60
$15.00	$13.80	$14.40	$14.80
$12.60	$12.60	$15.40	
$17.00	$15.60	$15.60	$14.80

Mean wage =

$13.74	$13.17	15.28	$12.24

The researcher wishes to determine if the apparent difference among the wage rates paid in these cities can be attributed to sampling error or is in fact a statistically significant finding.

In order to proceed to answer this research question, the researcher would conduct an analysis of variance test and begin by setting up a worksheet as shown in Worksheet 11.1.

Note that the individual city means differ from the overall mean and from each other. The total variation related to the overall distribution of wage rates can be expressed by the following equations:

Total sum of squares (SS_{Total}) = Between groups sum of squares $(SS_{BETWEEN})$
+ Within groups sum of squares (SS_{WITHIN})

This can be expressed in algebraic form as:

$$\sum f_x (x - \overline{x})^2 = \sum n_i (\overline{x}_i - \overline{x})^2 + \sum f_{x_i} (x_i - \overline{x}_i)^2 \qquad (11.6)$$

where
x = individual wage rates
x_i = individual wage rates by city
$\overline{x}$ = overall mean wage rate
$\overline{x}_i$ = mean wage rate by city
n_i = sample size for each city

WORKSHEET 11.1.   PUBLIC WORKS EMPLOYEES' EARNINGS.

X Hourly Wage	Overall f	X_1 Harrisburg f_1	X_2 Utica f_2	X_3 Annapolis f_3	X_4 Roanoke f_4
$10.40	3	1	2	0	0
$10.60	2	0	0	0	2
$12.60	6	3	1	0	2
$13.80	2	0	2	0	0
$14.40	1	0	0	1	0
$14.80	2	0	0	1	1
$15.00	2	1	0	1	0
$15.40	2	0	0	2	0
$15.60	4	0	2	2	0
$16.00	2	1	0	1	0
$17.00	1	1	0	0	0
	$n = 27$ $\bar{x} = \$13.77$	$n_1 = 7$ $\bar{x}_1 = \$13.74$	$n_2 = 7$ $\bar{x}_2 = \$13.17$	$n_3 = 8$ $\bar{x}_3 = \$15.28$	$n_4 = 5$ $\bar{x}_4 = \$12.24$

Calculation of Sum of Squares

Whenever equations contain expressions requiring the calculation of deviation, the arithmetic procedure becomes laborious. In order to minimize this problem in the case of analysis of variance, the researcher will find that such inconvenience can be minimized by calculating the SS_{TOTAL} and SS_{BETWEEN} and obtaining SS_{WITHIN} by subtraction ($SS_{\text{WITHIN}} = SS_{\text{TOTAL}} - SS_{\text{BETWEEN}}$).

In the example at hand, SS_{TOTAL} can be calculated in the following manner:

a.
$$SS_{\text{TOTAL}} = \sum f_x(x - \bar{x})^2 \tag{11.7}$$

$$3(10.40 - 13.77)^2 = 2(10.60 - 13.77)^2 + 6(12.60 - 13.77)^2$$
$$+ 2(13.80 - 13.77)^2 + 1(14.40 - 13.77)^2 + 2(14.80 - 13.77)^2$$
$$+ 2(15.00 - 13.77)^2 + 2(15.40 - 13.77)^2 = 4(15.60 - 13.77)^2$$
$$+ 2(16.00 - 13.77)^2 + 1(17.00 - 13.77)^2$$

$$SS_{\text{TOTAL}} = 107.24$$

b.

$$SS_{\text{BETWEEN}}$$
$$= \sum n_i(\bar{x}_i - \bar{x})^2 \tag{11.8}$$
$$= 7(13.74 - 13.77)^2 + 7(13.17 - 13.77)^2 + 8(15.28 - 13.77)^2 + 5(12.24 - 13.77)^2$$
$$= 32.47$$

c. $SS_{\text{WITHIN}} = SS_{\text{TOTAL}} - SS_{\text{BETWEEN}}$ (11.9)

$$= 107.24 - 32.47 = 74.77$$

Calculation of Variances and F Ratio

The essence of analysis of variance is the calculation of an F ratio that comprises the ratio of two variances. More specifically, it is the variance associated with SS_{BETWEEN} (S_b^2) divided by the variance associated the SS_{WITHIN} (S_W^2). When the sum of squares is divided by its appropriate degrees of freedom, the result is the variance, also known as the mean square (MS).

Hence,

$$MS_{\text{BETWEEN}} = \frac{SS_{\text{BETWEEN}}}{df_{\text{BETWEEN}}}$$ (11.10)

and

$$MS_{\text{WITHIN}} = \frac{SS_{\text{WITHIN}}}{df_{\text{WITHIN}}}$$ (11.11)

where
df_{between} = number of categories of the independent variable $(k) -1 = k - 1$
df_{within} = overall sample size minus the number of categories of the independent variable = $n - k$.

In terms of the problem at hand, therefore,

$$MS_{\text{BETWEEN}} = \frac{32.47}{3} = 10.82$$

$$MS_{\text{WITHIN}} = \frac{74.77}{23} = 3.25$$

The F ratio is calculated by dividing MS_{BETWEEN} by MS_{WITHIN}:

$$F = MS_{\text{BETWEEN}} \div MS_{\text{WITHIN}}$$ (11.12)

This process is best depicted in an ANOVA table. An ANOVA table (*AN*alysis *O*f *VA*riance) presents these calculations in an orderly, easily readable format. Figure 11.1 is the ANOVA table for this public works example, and it shows in one place the sum of squares and degrees of freedom for between, within, and total. It then divides these sums of squares by degrees of freedom and depicts the mean squares for between and within. Lastly, the ANOVA table divides MS_{BETWEEN} by MS_{WITHIN} to generate the F ratio.

FIGURE 11.1. ANOVA TABLE FOR PUBLIC WORKS EMPLOYEES.

Source of Variation	Sum of Squares	Degrees of Freedom	Mean Square	F
Between groups	32.47	$k - 1 = 3$	10.82	3.33
Within groups	74.77	$n - k = 23$	3.25	
Total	107.24	26		

Statistical Significance

In order to determine if the F ratio is statistically significant and therefore indicative of a genuine difference in wage rates among the cities, refer to the critical values of the F distribution in Exhibit 11.1 (95 percent confidence) and Exhibit 11.2 (99 percent confidence). The reader must determine from these exhibits a critical value of F that is identified by two factors: the confidence level (95 percent or 99 percent) and the degrees of freedom for both $MS_{BETWEEN}$ and MS_{WITHIN}. For this example, the critical F at 95 percent is 3.03, and at 99 percent, it is 4.77. The conclusion that the researcher can draw is that there is 95 percent confidence that the apparent difference in wage rates among these four cities is a genuine one and not sampling error. However, this conclusion cannot be drawn with 99 percent confidence.

Correlation Ratio E^2 and the Measure of Association—Eta

Eta is the appropriate measure of association when the independent variable is on the nominal scale and the dependent variable is on the interval scale. The equation for *Eta* is:

$$Eta = \sqrt{(SS_{BETWEEN} \div SS_{TOTAL})} \qquad (11.13)$$

The portion of the equation that is under the radical sign is known as the *correlation ratio* (E^2), and it represents the total variation explained by the grouping of the independent variable. In the example provided above, $E^2 = 32.47/107.24 = .303 \ldots$ and *Eta* $= .55$. Interpretation of the correlation ratio is that the city in which the worker is employed explains approximately 30 percent of the variation in pay, with all other factors accounting for the other 70 percent. Eta is interpreted in the context of Exhibit 10.2 (Cramér's V and phi). Hence, an Eta of .55 can be considered to be relatively strong.

EXHIBIT 11.1.   TABLE OF F-STATISTICS, P = 0.05.

df_NUM / df_DEN	1	2	3	4	5	6	7	8	9	10	11	12	13	14	15	16	17	18	19
3	10.13	9.55	9.28	9.12	9.01	8.94	8.89	8.85	8.81	8.79	8.76	8.74	8.73	8.71	8.70	8.69	8.68	8.67	8.67
4	7.71	6.94	6.59	6.39	6.26	6.16	6.09	6.04	6.00	5.96	5.94	5.91	5.89	5.87	5.86	5.84	5.83	5.82	5.81
5	6.61	5.79	5.41	5.19	5.05	4.95	4.88	4.82	4.77	4.74	4.70	4.68	4.66	4.64	4.62	4.60	4.59	4.58	4.57
6	5.99	5.14	4.76	4.53	4.39	4.28	4.21	4.15	4.10	4.06	4.03	4.00	3.98	3.96	3.94	3.92	3.91	3.90	3.88
7	5.59	4.74	4.35	4.12	3.97	3.87	3.79	3.73	3.68	3.64	3.60	3.57	3.55	3.53	3.51	3.49	3.48	3.47	3.46
8	5.32	4.46	4.07	3.84	3.69	3.58	3.50	3.44	3.39	3.35	3.31	3.28	3.26	3.24	3.22	3.20	3.19	3.17	3.16
9	5.12	4.26	3.86	3.63	3.48	3.37	3.29	3.23	3.18	3.14	3.10	3.07	3.05	3.03	3.01	2.99	2.97	2.96	2.95
10	4.96	4.10	3.71	3.48	3.33	3.22	3.14	3.07	3.02	2.98	2.94	2.91	2.89	2.86	2.85	2.83	2.81	2.80	2.79
11	4.84	3.98	3.59	3.36	3.20	3.09	3.01	2.95	2.90	2.85	2.82	2.79	2.76	2.74	2.72	2.70	2.69	2.67	2.66
12	4.75	3.89	3.49	3.26	3.11	3.00	2.91	2.85	2.80	2.75	2.72	2.69	2.66	2.64	2.62	2.60	2.58	2.57	2.56
13	4.67	3.81	3.41	3.18	3.03	2.92	2.83	2.77	2.71	2.67	2.63	2.60	2.58	2.55	2.53	2.51	2.50	2.48	2.47
14	4.60	3.74	3.34	3.11	2.96	2.85	2.76	2.70	2.65	2.60	2.57	2.53	2.51	2.48	2.46	2.44	2.43	2.41	2.40
15	4.54	3.68	3.29	3.06	2.90	2.79	2.71	2.64	2.59	2.54	2.51	2.48	2.45	2.42	2.40	2.38	2.37	2.35	2.34
16	4.49	3.63	3.24	3.01	2.85	2.74	2.66	2.59	2.54	2.49	2.46	2.42	2.40	2.37	2.35	2.33	2.32	2.30	2.29
17	4.45	3.59	3.20	2.96	2.81	2.70	2.61	2.55	2.49	2.45	2.41	2.38	2.35	2.33	2.31	2.29	2.27	2.26	2.24
18	4.41	3.55	3.16	2.93	2.77	2.66	2.58	2.51	2.46	2.41	2.37	2.34	2.31	2.29	2.27	2.25	2.23	2.22	2.20
19	4.38	3.52	3.13	2.90	2.74	2.63	2.54	2.48	2.42	2.38	2.34	2.31	2.28	2.26	2.23	2.21	2.20	2.18	2.17
20	4.35	3.49	3.10	2.87	2.71	2.60	2.51	2.45	2.39	2.35	2.31	2.28	2.25	2.23	2.20	2.18	2.17	2.15	2.14
22	4.30	3.44	3.05	2.82	2.66	2.55	2.46	2.40	2.34	2.30	2.26	2.23	2.20	2.17	2.15	2.13	2.11	2.10	2.08
24	4.26	3.40	3.01	2.78	2.62	2.51	2.42	2.36	2.30	2.25	2.22	2.18	2.15	2.13	2.11	2.09	2.07	2.05	2.04
26	4.23	3.37	2.98	2.74	2.59	2.47	2.39	2.32	2.27	2.22	2.18	2.15	2.12	2.09	2.07	2.05	2.03	2.02	2.00
28	4.20	3.34	2.95	2.71	2.56	2.45	2.36	2.29	2.24	2.19	2.15	2.12	2.09	2.06	2.04	2.02	2.00	1.99	1.97
30	4.17	3.32	2.92	2.69	2.53	2.42	2.33	2.27	2.21	2.16	2.13	2.09	2.06	2.04	2.01	1.99	1.98	1.96	1.95
35	4.12	3.27	2.87	2.64	2.49	2.37	2.29	2.22	2.16	2.11	2.08	2.04	2.01	1.99	1.96	1.94	1.92	1.91	1.89
40	4.08	3.23	2.84	2.61	2.45	2.34	2.25	2.18	2.12	2.08	2.04	2.00	1.97	1.95	1.92	1.90	1.89	1.87	1.85
45	4.06	3.20	2.81	2.58	2.42	2.31	2.22	2.15	2.10	2.05	2.01	1.97	1.94	1.92	1.89	1.87	1.86	1.84	1.82
50	4.03	3.18	2.79	2.56	2.40	2.29	2.20	2.13	2.07	2.03	1.99	1.95	1.92	1.89	1.87	1.85	1.83	1.81	1.80
60	4.00	3.15	2.76	2.53	2.37	2.25	2.17	2.10	2.04	1.99	1.95	1.92	1.89	1.86	1.84	1.82	1.80	1.78	1.76
70	3.98	3.13	2.74	2.50	2.35	2.23	2.14	2.07	2.02	1.97	1.93	1.89	1.86	1.84	1.81	1.79	1.77	1.75	1.74
80	3.96	3.11	2.72	2.49	2.33	2.21	2.13	2.06	2.00	1.95	1.91	1.88	1.84	1.82	1.79	1.77	1.75	1.73	1.72
100	3.94	3.09	2.70	2.46	2.31	2.19	2.10	2.03	1.97	1.93	1.89	1.85	1.82	1.79	1.77	1.75	1.73	1.71	1.69
200	3.89	3.04	2.65	2.42	2.26	2.14	2.06	1.98	1.93	1.88	1.84	1.80	1.77	1.74	1.72	1.69	1.67	1.66	1.64
500	3.86	3.01	2.62	2.39	2.23	2.12	2.03	1.96	1.90	1.85	1.81	1.77	1.74	1.71	1.69	1.66	1.64	1.62	1.61
1000	3.85	3.00	2.61	2.38	2.22	2.11	2.02	1.95	1.89	1.84	1.80	1.76	1.73	1.70	1.68	1.65	1.63	1.61	1.60
>1000	1.04	3.00	2.61	2.37	2.21	2.10	2.01	1.94	1.88	1.83	1.79	1.75	1.72	1.69	1.67	1.64	1.62	1.61	1.59

df_NUM / df_DEN	20	22	24	26	28	30	35	40	45	50	60	70	80	100	200	500	1,000	>1,000
3	8.66	8.65	8.64	8.63	8.62	8.62	8.60	8.59	8.59	8.58	8.57	8.57	8.56	8.55	8.54	8.53	8.53	8.54
4	5.80	5.79	5.77	5.76	5.75	5.75	5.73	5.72	5.71	5.70	5.69	5.68	5.67	5.66	5.65	5.64	5.63	5.63
5	4.56	4.54	4.53	4.52	4.50	4.50	4.48	4.46	4.45	4.44	4.43	4.42	4.41	4.39	4.37	4.37	4.36	4.36
6	3.87	3.86	3.84	3.83	3.82	3.81	3.79	3.77	3.76	3.75	3.74	3.73	3.72	3.71	3.69	3.68	3.67	3.67
7	3.44	3.43	3.41	3.40	3.39	3.38	3.36	3.34	3.33	3.32	3.30	3.29	3.29	3.27	3.25	3.24	3.23	3.23
8	3.15	3.13	3.12	3.10	3.09	3.08	3.06	3.04	3.03	3.02	3.01	2.99	2.99	2.97	2.95	2.94	2.93	2.93
9	2.94	2.92	2.90	2.89	2.87	2.86	2.84	2.83	2.81	2.80	2.79	2.78	2.77	2.76	2.73	2.72	2.71	2.71
10	2.77	2.75	2.74	2.72	2.71	2.70	2.68	2.66	2.65	2.64	2.62	2.61	2.60	2.59	2.56	2.55	2.54	2.54
11	2.65	2.63	2.61	2.59	2.58	2.57	2.55	2.53	2.52	2.51	2.49	2.48	2.47	2.46	2.43	2.42	2.41	2.41
12	2.54	2.52	2.51	2.49	2.48	2.47	2.44	2.43	2.41	2.40	2.38	2.37	2.36	2.35	2.32	2.31	2.30	2.30
13	2.46	2.44	2.42	2.41	2.39	2.38	2.36	2.34	2.33	2.31	2.30	2.28	2.27	2.26	2.23	2.22	2.21	2.21
14	2.39	2.37	2.35	2.33	2.32	2.31	2.28	2.27	2.25	2.24	2.22	2.21	2.20	2.19	2.16	2.14	2.14	2.13
15	2.33	2.31	2.29	2.27	2.26	2.25	2.22	2.20	2.19	2.18	2.16	2.15	2.14	2.12	2.10	2.08	2.07	2.07
16	2.28	2.25	2.24	2.22	2.21	2.19	2.17	2.15	2.14	2.12	2.11	2.09	2.08	2.07	2.04	2.02	2.02	2.01
17	2.23	2.21	2.19	2.17	2.16	2.15	2.12	2.10	2.09	2.08	2.06	2.05	2.03	2.02	1.99	1.97	1.97	1.96
18	2.19	2.17	2.15	2.13	2.12	2.11	2.08	2.06	2.05	2.04	2.02	2.00	1.99	1.98	1.95	1.93	1.92	1.92
19	2.16	2.13	2.11	2.10	2.08	2.07	2.05	2.03	2.01	2.00	1.98	1.97	1.96	1.94	1.91	1.89	1.88	1.88
20	2.12	2.10	2.08	2.07	2.05	2.04	2.01	1.99	1.98	1.97	1.95	1.93	1.92	1.91	1.88	1.86	1.85	1.84
22	2.07	2.05	2.03	2.01	2.00	1.98	1.96	1.94	1.92	1.91	1.89	1.88	1.86	1.85	1.82	1.80	1.79	1.78
24	2.03	2.00	1.98	1.97	1.95	1.94	1.91	1.89	1.88	1.86	1.84	1.83	1.82	1.80	1.77	1.75	1.74	1.73
26	1.99	1.97	1.95	1.93	1.91	1.90	1.87	1.85	1.84	1.82	1.80	1.79	1.78	1.76	1.73	1.71	1.70	1.69
28	1.96	1.93	1.91	1.90	1.88	1.87	1.84	1.82	1.80	1.79	1.77	1.75	1.74	1.73	1.69	1.67	1.66	1.66
30	1.93	1.91	1.89	1.87	1.85	1.84	1.81	1.79	1.77	1.76	1.74	1.72	1.71	1.70	1.66	1.64	1.63	1.62
35	1.88	1.85	1.83	1.82	1.80	1.79	1.76	1.74	1.72	1.70	1.68	1.66	1.65	1.63	1.60	1.57	1.57	1.56
40	1.84	1.81	1.79	1.77	1.76	1.74	1.72	1.69	1.67	1.66	1.64	1.62	1.61	1.59	1.55	1.53	1.52	1.51
45	1.81	1.78	1.76	1.74	1.73	1.71	1.68	1.66	1.64	1.63	1.60	1.59	1.57	1.55	1.51	1.49	1.48	1.47
50	1.78	1.76	1.74	1.72	1.70	1.69	1.66	1.63	1.61	1.60	1.58	1.56	1.54	1.52	1.48	1.46	1.45	1.44
60	1.75	1.72	1.70	1.68	1.66	1.65	1.62	1.59	1.57	1.56	1.53	1.52	1.50	1.48	1.44	1.41	1.40	1.39
70	1.72	1.70	1.67	1.65	1.64	1.62	1.59	1.57	1.55	1.53	1.50	1.49	1.47	1.45	1.40	1.37	1.36	1.35
80	1.70	1.68	1.65	1.63	1.62	1.60	1.57	1.54	1.52	1.51	1.48	1.46	1.45	1.43	1.38	1.35	1.34	1.33
100	1.68	1.65	1.63	1.61	1.59	1.57	1.54	1.52	1.49	1.48	1.45	1.43	1.41	1.39	1.34	1.31	1.30	1.28
200	1.62	1.60	1.57	1.55	1.53	1.52	1.48	1.46	1.43	1.41	1.39	1.36	1.35	1.32	1.26	1.22	1.21	1.19
500	1.59	1.56	1.54	1.52	1.50	1.48	1.45	1.42	1.40	1.38	1.35	1.32	1.30	1.28	1.21	1.16	1.14	1.12
1,000	1.58	1.55	1.53	1.51	1.49	1.47	1.43	1.41	1.38	1.36	1.33	1.31	1.29	1.26	1.19	1.13	1.11	1.08
>1,000	1.57	1.54	1.52	1.50	1.48	1.46	1.42	1.40	1.37	1.35	1.32	1.30	1.28	1.25	1.17	1.11	1.08	1.03

EXHIBIT 11.2.   TABLE OF F-STATISTICS, $P = 0.01$.

df_{NUM}/df_{DEN}	1	2	3	4	5	6	7	8	9	10	11	12	13	14	15
3	34.12	30.82	29.46	28.71	28.24	27.91	27.67	27.49	27.35	27.23	27.13	27.05	26.98	26.92	26.87
4	21.20	18.00	16.69	15.98	15.52	15.21	14.98	14.80	14.66	14.55	14.45	14.37	14.31	14.25	14.20
5	16.26	13.27	12.06	11.39	10.97	10.67	10.46	10.29	10.16	10.05	9.96	9.89	9.82	9.77	9.72
6	13.75	10.92	9.78	9.15	8.75	8.47	8.26	8.10	7.98	7.87	7.79	7.72	7.66	7.61	7.56
7	12.25	9.55	8.45	7.85	7.46	7.19	6.99	6.84	6.72	6.62	6.54	6.47	6.41	6.36	6.31
8	11.26	8.65	7.59	7.01	6.63	6.37	6.18	6.03	5.91	5.81	5.73	5.67	5.61	5.56	5.52
9	10.56	8.02	6.99	6.42	6.06	5.80	5.61	5.47	5.35	5.26	5.18	5.11	5.05	5.01	4.96
10	10.04	7.56	6.55	5.99	5.64	5.39	5.20	5.06	4.94	4.85	4.77	4.71	4.65	4.60	4.56
11	9.65	7.21	6.22	5.67	5.32	5.07	4.89	4.74	4.63	4.54	4.46	4.40	4.34	4.29	4.25
12	9.33	6.93	5.95	5.41	5.06	4.82	4.64	4.50	4.39	4.30	4.22	4.16	4.10	4.05	4.01
13	9.07	6.70	5.74	5.21	4.86	4.62	4.44	4.30	4.19	4.10	4.02	3.96	3.91	3.86	3.82
14	8.86	6.51	5.56	5.04	4.70	4.46	4.28	4.14	4.03	3.94	3.86	3.80	3.75	3.70	3.66
15	8.68	6.36	5.42	4.89	4.56	4.32	4.14	4.00	3.89	3.80	3.73	3.67	3.61	3.56	3.52
16	8.53	6.23	5.29	4.77	4.44	4.20	4.03	3.89	3.78	3.69	3.62	3.55	3.50	3.45	3.41
17	8.40	6.11	5.19	4.67	4.34	4.10	3.93	3.79	3.68	3.59	3.52	3.46	3.40	3.35	3.31
18	8.29	6.01	5.09	4.58	4.25	4.01	3.84	3.71	3.60	3.51	3.43	3.37	3.32	3.27	3.23
19	8.19	5.93	5.01	4.50	4.17	3.94	3.77	3.63	3.52	3.43	3.36	3.30	3.24	3.19	3.15
20	8.10	5.85	4.94	4.43	4.10	3.87	3.70	3.56	3.46	3.37	3.29	3.23	3.18	3.13	3.09
22	7.95	5.72	4.82	4.31	3.99	3.76	3.59	3.45	3.35	3.26	3.18	3.12	3.07	3.02	2.98
24	7.82	5.61	4.72	4.22	3.90	3.67	3.50	3.36	3.26	3.17	3.09	3.03	2.98	2.93	2.89
26	7.72	5.53	4.64	4.14	3.82	3.59	3.42	3.29	3.18	3.09	3.02	2.96	2.90	2.86	2.82
28	7.64	5.45	4.57	4.07	3.75	3.53	3.36	3.23	3.12	3.03	2.96	2.90	2.84	2.79	2.75
30	7.56	5.39	4.51	4.02	3.70	3.47	3.30	3.17	3.07	2.98	2.91	2.84	2.79	2.74	2.70
35	7.42	5.27	4.40	3.91	3.59	3.37	3.20	3.07	2.96	2.88	2.80	2.74	2.69	2.64	2.60
40	7.31	5.18	4.31	3.83	3.51	3.29	3.12	2.99	2.89	2.80	2.73	2.66	2.61	2.56	2.52
45	7.23	5.11	4.25	3.77	3.45	3.23	3.07	2.94	2.83	2.74	2.67	2.61	2.55	2.51	2.46
50	7.17	5.06	4.20	3.72	3.41	3.19	3.02	2.89	2.79	2.70	2.63	2.56	2.51	2.46	2.42
60	7.08	4.98	4.13	3.65	3.34	3.12	2.95	2.82	2.72	2.63	2.56	2.50	2.44	2.39	2.35
70	7.01	4.92	4.07	3.60	3.29	3.07	2.91	2.78	2.67	2.59	2.51	2.45	2.40	2.35	2.31
80	6.96	4.88	4.04	3.56	3.26	3.04	2.87	2.74	2.64	2.55	2.48	2.42	2.36	2.31	2.27
100	6.90	4.82	3.98	3.51	3.21	2.99	2.82	2.69	2.59	2.50	2.43	2.37	2.31	2.27	2.22
200	6.76	4.71	3.88	3.41	3.11	2.89	2.73	2.60	2.50	2.41	2.34	2.27	2.22	2.17	2.13
500	6.69	4.65	3.82	3.36	3.05	2.84	2.68	2.55	2.44	2.36	2.28	2.22	2.17	2.12	2.07
1000	6.66	4.63	3.80	3.34	3.04	2.82	2.66	2.53	2.43	2.34	2.27	2.20	2.15	2.10	2.06
>1000	1.04	4.61	3.78	3.32	3.02	2.80	2.64	2.51	2.41	2.32	2.25	2.19	2.13	2.08	2.04

df_{NUM}/df_{DEN}	16	17	18	19	20	22	24	26	28	30	35	40	45	50	60
3	26.83	26.79	26.75	26.72	26.69	26.64	26.60	26.56	26.53	26.50	26.45	26.41	26.38	26.35	26.32
4	14.15	14.11	14.08	14.05	14.02	13.97	13.93	13.89	13.86	13.84	13.79	13.75	13.71	13.69	13.65
5	9.68	9.64	9.61	9.58	9.55	9.51	9.47	9.43	9.40	9.38	9.33	9.29	9.26	9.24	9.20
6	7.52	7.48	7.45	7.42	7.40	7.35	7.31	7.28	7.25	7.23	7.18	7.14	7.11	7.09	7.06
7	6.28	6.24	6.21	6.18	6.16	6.11	6.07	6.04	6.02	5.99	5.94	5.91	5.88	5.86	5.82
8	5.48	5.44	5.41	5.38	5.36	5.32	5.28	5.25	5.22	5.20	5.15	5.12	5.09	5.07	5.03
9	4.92	4.89	4.86	4.83	4.81	4.77	4.73	4.70	4.67	4.65	4.60	4.57	4.54	4.52	4.48
10	4.52	4.49	4.46	4.43	4.41	4.06	4.33	4.30	4.27	4.25	4.20	4.17	4.14	4.12	4.08
11	4.21	4.18	4.15	4.12	4.10	4.06	4.02	3.99	3.96	3.94	3.89	3.86	3.83	3.81	3.78
12	3.97	3.94	3.91	3.88	3.86	3.82	3.78	3.75	3.72	3.70	3.65	3.62	3.59	3.57	3.54
13	3.78	3.75	3.72	3.69	3.66	3.62	3.59	3.56	3.53	3.51	3.46	3.43	3.40	3.38	3.34
14	3.62	3.59	3.56	3.53	3.51	3.46	3.43	3.40	3.37	3.35	3.30	3.27	3.24	3.22	3.18
15	3.49	3.45	3.42	3.40	3.37	3.33	3.29	3.26	3.24	3.21	3.17	3.13	3.10	3.08	3.05
16	3.37	3.34	3.31	3.28	3.26	3.22	3.18	3.15	3.12	3.10	3.05	3.02	2.99	2.97	2.93
17	3.27	3.24	3.21	3.19	3.16	3.12	3.08	3.05	3.03	3.00	2.96	2.92	2.89	2.87	2.83
18	3.19	3.16	3.13	3.10	3.08	3.03	3.00	2.97	2.94	2.92	2.87	2.84	2.81	2.78	2.75
19	3.12	3.08	3.05	3.03	3.00	2.96	2.92	2.89	2.87	2.84	2.80	2.76	2.73	2.71	2.67
20	3.05	3.02	2.99	2.96	2.94	2.90	2.86	2.83	2.80	2.78	2.73	2.69	2.67	2.64	2.61
22	2.94	2.91	2.88	2.85	2.83	2.78	2.75	2.72	2.69	2.67	2.62	2.58	2.55	2.53	2.50
24	2.85	2.82	2.79	2.76	2.74	2.70	2.66	2.63	2.60	2.58	2.53	2.49	2.46	2.44	2.40
26	2.78	2.75	2.72	2.69	2.66	2.62	2.58	2.55	2.53	2.50	2.45	2.42	2.39	2.36	2.33
28	2.72	2.68	2.65	2.63	2.60	2.56	2.52	2.49	2.46	2.44	2.39	2.35	2.32	2.30	2.26
30	2.66	2.63	2.60	2.57	2.55	2.51	2.47	2.44	2.41	2.39	2.34	2.30	2.27	2.25	2.21
35	2.56	2.53	2.50	2.47	2.44	2.40	2.36	2.33	2.31	2.28	2.23	2.19	2.16	2.14	2.10
40	2.48	2.45	2.42	2.39	2.37	2.33	2.29	2.26	2.23	2.20	2.15	2.11	2.08	2.06	2.02
45	2.43	2.39	2.36	2.34	2.31	2.27	2.23	2.20	2.17	2.14	2.09	2.05	2.02	2.00	1.96
50	2.38	2.35	2.32	2.29	2.27	2.22	2.18	2.15	2.12	2.10	2.05	2.01	1.97	1.95	1.91
60	2.31	2.28	2.25	2.22	2.20	2.15	2.12	2.08	2.05	2.03	1.98	1.94	1.90	1.88	1.84
70	2.27	2.23	2.20	2.18	2.15	2.11	2.07	2.03	2.01	1.98	1.93	1.89	1.85	1.83	1.78
80	2.23	2.20	2.17	2.14	2.12	2.07	2.03	2.00	1.97	1.94	1.89	1.85	1.82	1.79	1.75
100	2.19	2.15	2.12	2.09	2.07	2.02	1.98	1.95	1.92	1.89	1.84	1.80	1.76	1.74	1.69
200	2.09	2.06	2.03	2.00	1.97	1.93	1.89	1.85	1.82	1.79	1.74	1.69	1.66	1.63	1.58
500	2.04	2.00	1.97	1.94	1.92	1.87	1.83	1.79	1.76	1.74	1.68	1.63	1.60	1.57	1.52
1000	2.02	1.98	1.95	1.92	1.90	1.85	1.81	1.77	1.74	1.72	1.66	1.61	1.58	1.54	1.50
>1000	2.00	1.97	1.94	1.91	1.88	1.83	1.79	1.76	1.73	1.70	1.64	1.59	1.56	1.53	1.48

Post Hoc Tests: A Multiple Comparison of Means

A significant F ratio informs the researcher that an overall difference among groups exists. If only two groups were being investigated, no further analysis would be required. However, when a significant F is obtained for three or more groups, this does not mean that all groups differ from all others—just that at least one is different from one other. In such cases, it is important to determine precisely which groups are significantly different from the others and which are not.

Repeated applications of independent or paired samples t tests for two groups would increase the chance that a Type I error might occur. Fortunately there exist a number of statistical significance tests, known as *post hoc tests,* that have been developed to make multiple comparisons of means after a significant F has been found. These tests can pinpoint where the significant means differences exist.

One of the most widely used post hoc tests is Tukey's HSD (Honestly Significant Difference). Tukey's HSD is used only after a significant F has been obtained. Tukey compares the difference between any two means against the calculated HSD, and a difference between any two means is significant only if it equals or exceeds the HSD. The equation for HSD is as follows:

$$HSD = q\sqrt{(MS_{\text{WITHIN}} \div \overline{x}_h)} \tag{11.14}$$

where

q = the Studentized range value from Exhibit 11.3 that corresponds to the confidence level and number of groups (k)[1]

$\overline{x}_h$ = harmonic mean of group sizes, which is calculated as the kth root of the product of the sample sizes or

$$\sqrt[k]{(n_1)(n_2)(n_3)\ldots(n_k)} \tag{11.15}$$

To illustrate HSD from the example of the public works employees,

$q = 3.915$ (95 percent confidence—df for $MS_{\text{WITHIN}} = 23 -$ number of groups $= 4$). The reader will note that $df = 23$ is not contained on Exhibit 11.3. The q value of 3.915 has been interpolated between df $=20$ and df $= 24$.

$$\overline{x}_h = \sqrt[4]{(7)(7)(8)(5)} = 6.654$$
$$HSD = 3.915\left(\sqrt{(3.25 \div 6.654)}\right)$$
$$= 3.915(.699)$$
$$= \$2.74$$

EXHIBIT 11.3. STUDENTIZED RANGE (Q) DISTRIBUTION.

Number of Groups (K)

v	α	2	3	4	5	6	7	8	9	10	11	12	13	14	15	16	17	18	19	20
1	.050	17.97	26.98	32.82	37.08	40.41	43.12	45.40	47.36	49.07	50.59	51.96	53.19	54.32	55.36	56.32	57.21	58.04	58.82	59.55
1	.010	89.98	135.0	164.3	185.6	202.2	215.7	227.1	236.9	245.5	253.1	259.9	266.1	271.8	276.9	281.7	286.2	290.3	294.2	297.9
2	.050	6.085	8.331	9.798	10.88	11.73	12.43	13.03	13.54	13.99	14.39	14.75	15.08	15.37	15.65	15.91	16.14	16.36	16.57	16.77
2	.010	14.03	19.02	22.29	24.72	26.63	28.20	29.53	30.68	31.69	32.59	33.39	34.13	34.80	35.42	35.99	36.53	37.03	37.50	37.94
3	.050	4.501	5.910	6.825	7.502	8.037	8.478	8.852	9.177	9.462	9.717	9.946	10.15	10.35	10.52	10.69	10.84	10.98	11.11	11.24
3	.010	8.260	10.62	12.17	13.32	14.24	15.00	15.64	16.20	16.69	17.13	17.52	17.88	18.21	18.52	18.80	19.06	19.31	19.54	19.76
4	.050	3.927	5.040	5.757	6.287	6.706	7.053	7.347	7.602	7.826	8.027	8.208	8.373	8.524	8.664	8.793	8.914	9.027	9.133	9.233
4	.010	6.511	8.120	9.173	9.958	10.58	11.10	11.54	11.93	12.26	12.57	12.84	13.09	13.32	13.53	13.73	13.91	14.08	14.24	14.39
5	.050	3.635	4.602	5.218	5.673	6.033	6.330	6.582	6.802	6.995	7.167	7.324	7.465	7.596	7.716	7.828	7.932	8.030	8.122	8.208
5	.010	5.702	6.976	7.806	8.421	8.913	9.321	9.669	9.971	10.24	10.48	10.70	10.89	11.08	11.24	11.40	11.54	11.68	11.81	11.93
6	.050	3.460	4.339	4.896	5.305	5.628	5.895	6.122	6.319	6.493	6.649	6.789	6.917	7.034	7.143	7.244	7.338	7.426	7.509	7.587
6	.010	5.243	6.331	7.033	7.556	7.974	8.318	8.611	8.869	9.097	9.300	9.485	9.653	9.808	9.951	10.08	10.21	10.32	10.43	10.54
7	.050	3.344	4.165	4.681	5.060	5.359	5.606	5.815	5.997	6.158	6.302	6.431	6.550	6.658	6.759	6.852	6.939	7.020	7.097	7.169
7	.010	4.948	5.919	6.543	7.006	7.373	7.678	7.940	8.167	8.368	8.548	8.711	8.859	8.996	9.124	9.242	9.353	9.456	9.553	9.645
8	.050	3.261	4.041	4.529	4.886	5.167	5.399	5.596	5.767	5.918	6.053	6.175	6.287	6.389	6.483	6.571	6.653	6.729	6.801	6.870
8	.010	4.745	5.635	6.204	6.625	6.960	7.238	7.475	7.681	7.864	8.028	8.176	8.312	8.437	8.552	8.659	8.760	8.854	8.942	9.026
9	.050	3.199	3.948	4.415	4.755	5.024	5.244	5.432	5.595	5.738	5.867	5.983	6.089	6.186	6.276	6.359	6.437	6.510	6.579	6.644
9	.010	4.595	5.428	5.957	6.347	6.658	6.915	7.134	7.326	7.495	7.647	7.785	7.910	8.026	8.133	8.233	8.326	8.413	8.495	8.573
10	.05	3.151	3.877	4.327	4.654	4.912	5.124	5.304	5.460	5.598	5.722	5.833	5.935	6.028	6.114	6.194	6.269	6.339	6.405	6.467
10	.01	4.482	5.270	5.769	6.136	6.428	6.669	6.875	7.055	7.214	7.356	7.485	7.603	7.712	7.813	7.906	7.994	8.076	8.153	8.226
11	.05	3.113	3.820	4.256	4.574	4.823	5.028	5.202	5.353	5.486	5.605	5.713	5.811	5.901	5.984	6.062	6.134	6.202	6.265	6.325
11	.01	4.392	5.146	5.621	5.970	6.247	6.476	6.672	6.842	6.992	7.127	7.250	7.362	7.465	7.560	7.649	7.732	7.809	7.883	7.952
12	.05	3.081	3.773	4.199	4.508	4.748	4.947	5.116	5.262	5.395	5.510	5.615	5.710	5.797	5.878	5.953	6.023	6.089	6.151	6.209
12	.01	4.320	5.046	5.502	5.836	6.101	6.321	6.507	6.670	6.814	6.943	7.060	7.167	7.265	7.356	7.441	7.520	7.594	7.664	7.731
13	.05	3.055	3.734	4.151	4.453	4.690	4.884	5.049	5.192	5.318	5.431	5.533	5.625	5.711	5.789	5.862	5.930	5.994	6.055	6.112
13	.01	4.260	4.964	5.404	5.727	5.981	6.192	6.372	6.528	6.670	6.791	6.903	7.006	7.100	7.188	7.269	7.345	7.417	7.484	7.548
14	.05	3.033	3.701	4.111	4.407	4.639	4.829	4.990	5.130	5.253	5.363	5.463	5.554	5.637	5.714	5.785	5.852	5.915	5.973	6.029
14	.01	4.210	4.895	5.322	5.634	5.881	6.085	6.258	6.409	6.546	6.664	6.772	6.871	6.962	7.047	7.125	7.199	7.268	7.333	7.394
15	.05	3.014	3.673	4.076	4.367	4.595	4.782	4.940	5.077	5.198	5.306	5.403	5.492	5.574	5.649	5.719	5.785	5.846	5.904	5.958
15	.01	4.167	4.836	5.252	5.556	5.796	5.994	6.162	6.309	6.438	6.555	6.660	6.757	6.845	6.927	7.003	7.074	7.141	7.204	7.264
16	.05	2.998	3.649	4.046	4.333	4.557	4.741	4.896	5.031	5.150	5.256	5.352	5.439	5.519	5.593	5.662	5.726	5.785	5.843	5.896
16	.01	4.131	4.786	5.192	5.488	5.722	5.915	6.079	6.222	6.348	6.461	6.564	6.658	6.743	6.824	6.897	6.967	7.032	7.093	7.151
17	.05	2.984	3.628	4.020	4.303	4.524	4.705	4.858	4.991	5.108	5.212	5.306	5.392	5.471	5.544	5.612	5.675	5.734	5.790	5.842
17	.01	4.099	4.742	5.140	5.430	5.659	5.847	6.007	6.147	6.270	6.380	6.480	6.572	6.656	6.733	6.806	6.873	6.937	6.997	7.053
18	.05	2.971	3.609	3.997	4.276	4.494	4.673	4.824	4.955	5.071	5.173	5.266	5.351	5.429	5.501	5.567	5.629	5.688	5.743	5.794
18	.01	4.071	4.703	5.094	5.379	5.603	5.787	5.944	6.081	6.201	6.309	6.407	6.496	6.579	6.655	6.725	6.791	6.854	6.912	6.967
19	.05	2.960	3.593	3.977	4.253	4.468	4.645	4.794	4.924	5.037	5.139	5.231	5.314	5.391	5.462	5.528	5.589	5.647	5.701	5.752
19	.01	4.046	4.669	5.054	5.333	5.553	5.735	5.888	6.022	6.141	6.246	6.342	6.430	6.510	6.585	6.654	6.719	6.780	6.837	6.891
20	.05	2.950	3.578	3.958	4.232	4.445	4.620	4.768	4.895	5.008	5.108	5.199	5.282	5.357	5.427	5.492	5.553	5.610	5.663	5.714
20	.01	4.024	4.639	5.018	5.293	5.509	5.687	5.839	5.970	6.086	6.190	6.285	6.370	6.449	6.523	6.591	6.654	6.714	6.770	6.823
24	.05	2.919	3.523	3.890	4.153	4.358	4.526	4.667	4.789	4.897	4.993	5.079	5.158	5.231	5.298	5.360	5.418	5.472	5.522	5.570
24	.01	3.942	4.527	4.884	5.143	5.346	5.513	5.654	5.777	5.885	5.982	6.070	6.150	6.223	6.291	6.355	6.414	6.469	6.521	6.571
30	.05	2.888	3.487	3.845	4.102	4.301	4.464	4.601	4.720	4.824	4.917	5.001	5.077	5.147	5.211	5.271	5.327	5.379	5.429	5.475
30	.01	3.889	4.454	4.799	5.048	5.242	5.401	5.536	5.653	5.756	5.848	5.932	6.008	6.078	6.143	6.203	6.259	6.311	6.360	6.407
40	.05	2.858	3.442	3.791	4.039	4.232	4.388	4.521	4.634	4.735	4.824	4.904	4.977	5.044	5.106	5.163	5.216	5.266	5.313	5.358
40	.01	3.825	4.367	4.695	4.931	5.114	5.265	5.392	5.502	5.599	5.685	5.764	5.835	5.900	5.961	6.017	6.069	6.118	6.165	6.208
50	.05	2.841	3.416	3.758	4.002	4.190	4.344	4.473	4.584	4.681	4.768	4.847	4.918	4.983	5.043	5.098	5.150	5.199	5.245	5.288
50	.01	3.787	4.316	4.634	4.863	5.040	5.185	5.308	5.414	5.507	5.590	5.665	5.734	5.796	5.854	5.908	5.958	6.005	6.050	6.092
60	.05	2.829	3.399	3.737	3.977	4.163	4.314	4.441	4.550	4.646	4.732	4.808	4.878	4.942	5.001	5.056	5.107	5.154	5.199	5.241
60	.01	3.762	4.282	4.594	4.818	4.991	5.133	5.253	5.356	5.447	5.528	5.601	5.667	5.728	5.784	5.837	5.886	5.931	5.974	6.015
80	.05	2.814	3.377	3.711	3.947	4.129	4.278	4.402	4.509	4.603	4.686	4.761	4.829	4.892	4.949	5.003	5.052	5.099	5.142	5.183
80	.01	3.732	4.241	4.545	4.763	4.931	5.069	5.185	5.284	5.372	5.451	5.521	5.585	5.644	5.698	5.749	5.796	5.840	5.881	5.920
100	.05	2.806	3.365	3.695	3.929	4.109	4.256	4.379	4.484	4.577	4.659	4.733	4.800	4.862	4.918	4.971	5.020	5.066	5.108	5.149
100	.01	3.714	4.216	4.516	4.730	4.896	5.031	5.144	5.242	5.328	5.405	5.474	5.537	5.594	5.648	5.698	5.743	5.786	5.826	5.864
> 100	.05	2.800	3.356	3.685	3.917	4.096	4.241	4.363	4.468	4.560	4.641	4.714	4.781	4.842	4.898	4.950	4.998	5.043	5.086	5.126
> 100	.01	3.702	4.200	4.497	4.709	4.872	5.005	5.118	5.214	5.299	5.375	5.443	5.505	5.561	5.614	5.662	5.708	5.750	5.790	5.827
> 120	.05	2.77	3.31	3.63	3.86	4.03	4.17	4.29	4.39	4.47	4.55	4.62	4.69	4.75	4.81	4.86	4.93	4.98	5.02	5.06
> 120	.01	3.64	4.12	4.40	4.60	4.76	4.88	4.99	5.08	5.16	5.23	5.30	5.36	5.42	5.48	5.54	5.59	5.64	5.69	5.74

df within

If the difference between the mean wages in any two cities equals or exceeds $2.74, then those two cities have wage structures for public works employees that are statistically different from each other. In this example, there is one such difference—Annapolis and Roanoke, which differ by $3.04 ($15.28 − $12.24). The other combinations of cities cannot be considered to differ from each other.

Spuriousness

Regarding the significance tests shown in Chapters Seven, Ten, and this chapter, it should be made clear that it is sometimes incorrect to draw conclusions based on the apparent relationship between only two variables. The researcher must account for the possibility that factors other than a single independent variable may also have some influence on the dependent variable and that even if these factors are very evident in the data, their effect may be obscured if they are excluded from the analysis. For example, suppose that it is found through a survey research study that football fans earn a statistically significant higher annual income than non-fans do. Prior to making a policy decision based on such a finding, the researcher should analyze other variables that might be suspected of influencing the independent variable (football fans) and the dependent variable (annual income). One such variable would be the gender of the respondent because football fans tend to be predominantly male and income levels are higher among males. It may therefore be that the preliminary finding is erroneous and that the actual independent variable is gender. If this should be the case, then the relationship between football fans and income is said to be *spurious*, or not genuine.

There are a variety of advanced statistical techniques that are designed to measure the influence of more than one independent variable on a dependent variable and thereby identify spurious relationships. Foremost among these techniques are multiple regression analysis and partial correlation, both of which are introduced briefly in Chapter Twelve but are otherwise beyond the scope of this text.

EXERCISES

1. A sample of 60 men and 50 women revealed that the men are 65 percent in favor of district elections for city council seats, but only 41 percent of the women support it. Test the data to determine if men are significantly more in favor of district elections than are women or if the apparent finding is only sampling error. Use the 99 percent confidence level.

2. A survey of 500 Democrats finds that Democrats have a mean age of 43.7 ($s = 8.6$). Statistics from the registrar of voters indicate that the average

Republican voter's mean age is 42.3 ($s = 7.8$). Republican data are based on a 200-person survey. Are Democrats significantly different in age from Republicans (99 percent confidence)?

3. A county welfare department wishes to study the magnitude of welfare stipends granted to needy families. A sample of 42 recipients is pulled randomly from the files (31 are white families, and 11 are black). You study the data and find that the white sample's mean payment is $419 and the black sample's mean is $381. The white sample's standard deviation is $60, and the black sample's standard deviation is $50. Can you conclude with 95 percent confidence that there is a real difference between welfare payments to whites and blacks, or is the apparent difference only sampling error?

4. Once upon a time, there were 152 bears of sufficiently variable characteristics to represent a random sample of bears. Some of these bears loved chocolate and others loved french fries. Among the 126 that were chocolate eaters, the mean number of calories they consumed each day from their chocolate diet was 54,865 ($s = 5,000$). The 26 french fry eaters averaged 57,305 calories from their diet ($s = 7,000$). Are these data sufficient to establish at 95 percent confidence that french fry–eating bears consume more calories in a day than do chocolate-eating bears?

5. Among 160 city planners surveyed, 46.0 percent disliked their job. Among 280 city engineers, only 37.0 percent disliked their job. Do planners dislike their jobs more than do engineers (99 percent level of confidence)?

6. City manager William "Blue Pencil" Pennypincher is concerned about low efficiency scores that his staff received at the state testing institute. Pennypincher believes that these scores are the result of his staff's being overcompensated and growing lazy. Pennypincher orders 75 randomly selected staff members to deposit 25 percent of their pay into the city's pension fund and to earn back the foregone 25 percent in incentives based on output. After 60 days, these staff members are tested again. The results are shown in Table 11.2 for the 75 participants. It appears as if Pennypincher's experiment has failed. Has it (95 percent confidence)?

7. General R. U. Coughin is concerned that his troops on leave are being too casual about contact with females. He has created a scale to measure the sexually transmitted disease score (STDS) for his troops. At a routine inspection of 125 soldiers, STDS $= 50.0$ ($s = 15$). Coughin orders the troops to watch films about protecting themselves and to reduce the amount of time

TABLE 11.2. EFFICIENCY TEST: CITY WORKERS.

	Original Test	Follow-Up Test
Mean	78.5	71.6
Standard deviation	30.3	26.2

they spend in town. After 3 months, he orders another inspection and finds that out of 200 troops, many of whom were in the first study, the STDS are 47.4 (s = 8). What can you say about the general's steps to alleviate the problem?

8. In order to comply with local air quality regulations, a small business sought to find out how far its employees commute to work each day. The survey of 25 employees showed a mean commute of 11 miles. As an interesting side issue, the company wanted to find out if the commute patterns differed by category of employee. Not knowing how to do this, the company has asked you to address this question with 95 percent confidence in your conclusions. The company supplied you with the following commute distances:

- Management (5 employees): 6, 7, 12, 3, 7
- Clerical/secretarial (8 employees): 10, 11, 5, 10, 11, 7, 7, 11
- Labor (12 employees): 11, 12, 14, 5, 20, 12, 10, 10, 12, 20, 30, 12

a. Does the commute mileage differ by employee category at 95 percent confidence?

b. How much of the total variation in commute mileage is explained by these categories?

c. If a significant difference exists, what employee category differences lead you to that conclusion?

9. Researchers traveled to 5 cities in the Imperial Valley to ask residents about their shopping patterns. Respondents were asked how many days per two-month period they shopped for goods and services outside the Valley. Thirty persons (indicated below under the cities in which they reside) participated in the study. They identified the number of these "outside" shopping days as follows:

El Centro	Calexico	Brawley	Imperial	Holtville
1	3	1	2	5
3	3	6	5	3
2	4	2	7	2
1	3	5	4	9
1	9	6	8	8
0	7	3	1	6

a. Does there appear to be a difference between the residents of these cities?

b. Based on these responses, determine whether these differences, if any, between the cities can be concluded to be statistically significant at the 99 percent confidence level.

c. Which cities have differences that have led to the significant difference (again, if any)?

d. Identify the strength of the relationship between city of residence and shopping outside the Valley.

10. A sample of 350 respondents was asked to indicate their favorite vacation destination from among Hawaii, Las Vegas, and Cabo San Lucas. The mean age of the 125 who favored Hawaii was 38.2. Of the 150 who chose Las Vegas, the mean age was 47.3. The remaining respondents favored Cabo San Lucas, and their mean age was 42.7. Total sum of squares for these data equals 100,000. Is there a statistically significant difference among these resort choices by age of visitor? The local chambers of commerce are awaiting your findings and ask that you be 95 percent confident in your conclusions. If you find a significant difference, which resort(s) differs significantly from the others?

Note

1. The Tukey test revolves around a measure known as the Studentized range statistic (q). For any particular pair of means among the k groups, let us designate the larger and smaller as M_L and M_S, respectively. The Studentized range statistic can then be calculated for any particular pair as

$$q = (M_L - M_S) \frac{\sqrt{MS_{WITHIN}}}{N_{p/s}} \quad \text{where } N_{p/s} \text{ is the number of values of } x_i.$$

CHAPTER TWELVE

REGRESSION AND CORRELATION

This chapter will consider the relationship between two interval scale variables. With interval variables as both independent and dependent, the researcher is not only able to perform tests of statistical significance and measures of association but also to quantify the exact relationship with mathematical precision through the designation of a line or curve that represents the relationship. When the researcher's interest is focused on the existence or strength of a relationship between two interval variables, tests of significance and measures of association may suffice. However, once such a relationship is found, the researcher will likely wish to predict the exact value of the dependent variable based on the value of the independent variable(s) through the application of regression analysis.

Simple Linear Regression

Interval scale variables afford the researcher significantly more informative data. For example, in the case of Figure 12.1, the government funding agency may ask the researcher a very practical question concerning the impact of specific subsidy levels. The agency might inquire how properties on a given city block can be expected to increase in value when total rehabilitation subsidies of $118,000

FIGURE 12.1.   REHABILITATION SUBSIDY/MARKET VALUE INCREASE ANALYSIS.

Block	Rehabilitation Subsidy (in thousands)	Mean Market Value Increase (in thousands)
A	100	3
B	85	3
C	95	1
D	130	5
E	110	5
F	75	2
G	135	4
H	60	0
I	120	4
J	115	3

are made available. This question can be answered by applying the technique of simple linear regression to the raw data in Figure 12.1.

Simple linear regression is performed on two interval variables. Simple linear regression analysis requires that the researcher once again identify the independent and dependent variables, using the principles discussed in Chapter Ten. In this example, the independent variable is rehabilitation subsidy (commonly designed as x), and the dependent variable is property value change (commonly designated as y). The objective of linear regression analysis is to establish the equation for the line that most closely aligns with the raw data. It is this line (the regression line) that permits the researcher to predict values of y based on values of x, thereby answering the question posed by the government funding agency. Figure 12.2 depicts the data from Figure 12.1 as plotted by its values of x and y. The regression line, or *line of best fit*, is the line that minimizes the sum of the perpendicular distances from each of the plotted points to that line. One and only one line meets this criterion. The regression line is also shown in Figure 12.2.

The equation for the regression line conforms to the general equation for a line and can be expressed as

$$y = a + bx \tag{12.1}$$

FIGURE 12.2. SCATTER PLOT AND REGRESSION LINE.

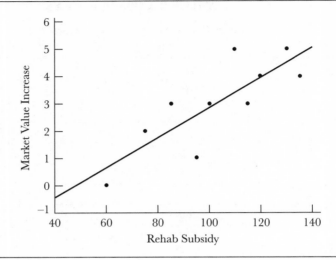

where

  $a =$ the y-intercept (the point at which the line crosses the y-axis)
  $b =$ the slope of the line (change in y divided by change in x)—also known
    as the regression coefficient.

Linear regression provides for the determination of the $y -$ intercept a and slope b of the line of best fit directly from raw data. The formulas for these calculations are as follows and are operationalized in Worksheet 12.1.

$$b = \frac{\sum (x - \overline{x})(y - \overline{y})}{\sum (x - \overline{x})^2} \tag{12.2}$$

$$a = \overline{y} - b(\overline{x}) \tag{12.3}$$

The regression line therefore is $y = -2.5275 + .051x$. This means that as subsidies increase by \$1,000, value increases by approximately \$50. The question posed by the agency concerning its \$118,000 subsidy can now be answered by the researcher as follows:

$$y = -2.5275 + .051(118)$$
$$= 3.4905$$

WORKSHEET 12.1. CALCULATION OF REGRESSION LINE.

Block	x	y	$x - \bar{x}$	$y - \bar{y}$	$(x - \bar{x})(y - \bar{y})$	$(x - \bar{x})^2$	$(y - \bar{y})^2$
A	100	3	−2.5	.3	−.75	6.25	.09
B	85	3	−17.5	.3	−5.25	306.25	.09
C	95	1	−7.5	−1.7	12.75	56.25	2.89
D	130	5	27.5	2.3	63.25	756.25	5.29
E	110	2	7.5	−.7	−5.25	56.25	.49
F	75	2	−27.5	−.7	19.25	756.25	.49
G	135	4	32.5	1.3	42.25	1056.25	1.69
H	60	0	−42.5	−2.7	114.75	1,806.25	7.29
I	120	4	17.5	1.3	22.75	306.25	1.69
J	115	3	12.5	.3	3.75	156.25	.09

$\bar{x} = 102.5$ $\bar{y} = 2.7$ $n = 10$

$b = \dfrac{267.5}{5262.5}$

$= .051$

Σ = sum of the products = SP = 267.5

Σ = sum of squares (x) = SS_x = 5,262.5

Σ = sum of squares (y) = SS_y = 20.1

$a = \bar{y} - b\bar{x}$

$= 2.7 - .051(102.5)$

$= 2.7 - 5.2275$

$= -2.5275$

The estimated mean property value increase in the community is projected to be $3,491.

It is important to note one of the crucial limitations on use of the line in this manner. If the subsidy were $40,000, for instance, the line would predict that property values would decline by approximately $500. There is no evidence that this is a logical and appropriate conclusion inasmuch as the data depict subsidies ranging from $60,000 to $130,000. It is not known what would happen to values when the projection is drawn for values far outside the data range. Regression lines are therefore potentially useful only within or slightly outside the range of independent variable data used to determine the line. Making predictions for independent variable values that may be far outside that range is not appropriate.

Testing the Significance of the Regression Line

The equations for the *y*-intercept and the slope can be calculated from any set of raw data. Hence, a regression equation can be established whether or not a significant linear relationship exists between the two interval scale variables. For

FIGURE 12.3.   VIOLENT CRIMES AND POPULATION.

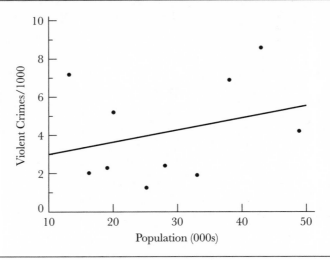

example, data points that are widely scattered, as in Figure 12.3, will yield a regression line, just as will those that closely approximate the line, as in Figure 12.4. Linear regression analysis therefore requires that all calculated regression lines be tested for statistical significance. If the line is found to be statistically significant, it may be considered for use for predictive purposes, subject to one further consideration, which we address following the test for statistical significance. If the regression line is not statistically significant, then it cannot be used for predictive purposes. Generally, under similar sampling conditions and sample sizes, the closer the data conform to the line of best fit, the more likely the line is to pass the test of statistical significance.

There are several alternative tests to determine whether the regression equation is significant. The most widely used test is an ANOVA (analysis of variance) test, which has been previously discussed in the context of determining the amount of the variance that the categories of the independent variable explain in comparison to the amount that remains as residual, unexplained variance within those categories (see Analysis of Variance in Chapter Eleven).

The ANOVA test for the significance of the regression line is similar to the analysis of variance test in that it compares the amount that the total variance is explained by using the regression line for predictive purposes rather than using

FIGURE 12.4. PRIOR CONVICTIONS AND SENTENCE LENGTH.

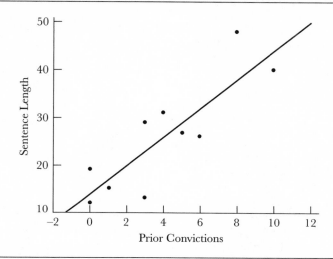

the mean. The total variance in the dependent variable using the mean for prediction is known as the *total mean square* (MS_{total})—just as it was in the analysis of variance test, and just as with analysis of variance, MS_{total} consists of two components—*residual mean square* ($MS_{residual}$) and *regression mean square* ($MS_{regression}$). $MS_{residual}$ is the variance that remains after using the regression line, and $MS_{regression}$ is the amount that the regression line reduces MS_{total} to $MS_{residual}$.

where

$$MS_{regression} = SS_{regression} \div df_{regression} \qquad (12.4)$$

$$MS_{residual} = SS_{residual} \div df_{residual} \qquad (12.5)$$

$$MS_{total} = SS_{total} \div df_{total} \qquad (12.6)$$

$$df_{total} = n - 1$$

$$df_{regression} = 1 \quad \text{(one for each independent variable)}$$

$$df_{residual} = n - 2$$

Calculation of Sum of Squares

Determination of the mean squares therefore requires, as an initial step, the determination of sums of squares, as follows:

$$SS_{\text{regression}} = \frac{[\Sigma(x - \bar{x})(y - \bar{y})]^2}{\Sigma(x - \bar{x})^2} \tag{12.7}$$

$$SS_{\text{residual}} = \Sigma(y - \hat{y})^2 \tag{12.8}$$

$$SS_{\text{total}} = \Sigma(y - \bar{y})^2 \tag{12.9}$$

As was the case with analysis of variance, whenever equations contain expressions requiring the calculation of deviation, the arithmetic procedure becomes laborious. In order to minimize this problem in the case of the ANOVA calculation for the regression line, the researcher will find that such inconvenience can be minimized by calculating SS_{total} and $SS_{\text{regression}}$ and obtaining SS_{residual} by subtraction ($SS_{\text{residual}} = SS_{\text{total}} - SS_{\text{regression}}$). This is the case because SS_{total} is found in the worksheet for determining the line, so it exists without additional calculations:

$$SS_{\text{total}} = SS_y \tag{12.10}$$

and because the equation for $SS_{\text{regression}}$ can be simplified to

$$SS_{\text{regression}} = b(\text{SP}) \tag{12.11}$$

or the slope of the line multiplied by the numerator of the slope calculation.

Were SS_{residual} to be calculated directly, it would involve another factor, $\hat{y}$, which is the predicted value of $\hat{y}$ using the regression line. In the case of the subsidy example, $\hat{y}$ is calculated by applying the regression line equation ($\hat{y} = -2.5275 + .051x$) to each value of x in the sample and then determining the squared differences with the actual $\hat{y}$ values, as shown in Worksheet 12.2.

In contrast to Worksheet 12.2, SS_{residual} can be calculated more simply by subtracting SS_{residual} from SS_{total}, as follows:

$$SS_{\text{total}} = SS_y = 20.1 \quad \text{(from Worksheet 12.1)}$$
$$SS_{\text{regression}} = b(\text{SP}) = .051(267.5) = 13.6425$$
$$SS_{\text{residual}} = 20.1 - 13.6425 = 6.4575 \quad \text{(difference from worksheet is in rounding } b = .0508 \text{ to } .051)$$

WORKSHEET 12.2.   RESIDUAL SUM OF SQUARES.

Block	x	y	$\hat{y}$	$y - \hat{y}$	$(y - \hat{y})^2$
A	100	3	2.5725	.4275	.1828
B	85	3	1.8075	1.1925	1.4221
C	95	1	2.3175	−1.3175	1.7358
D	130	5	4.1025	.8975	.8055
E	110	2	3.0825	−1.0825	1.1718
F	75	2	1.2975	.7025	.4935
G	135	4	4.3575	−.3575	.1278
H	60	0	.5325	−.5325	.2836
I	120	4	3.5925	.4075	.1661
J	115	3	3.3375	−.3375	.1139

$$\Sigma = SS_{residual} = 6.5029$$

The considerably greater ease with which $SS_{residual}$ is calculated in the subtraction method versus the calculations in Worksheet 12.2 should be obvious to the reader.

Calculation of Variances and F Ratio

Once the sums of squares are calculated, an ANOVA table can be constructed (Figure 12.5). As with analysis of variance, the ANOVA table compiles and simplifies the presentation of the data that are necessary to determine the significance of the regression line.

In referring to the critical values of the F distribution in Chapter Eleven (Exhibit 11.1 for 95 percent confidence and Exhibit 11.2 for 99 percent confidence), the researcher would find that with df numerator = 1 and df denominator = 8, the critical F at 95 percent is 5.32, and at 99 percent, it is 11.3. The

FIGURE 12.5.   ANOVA TABLE FOR REGRESSION LINE OF SUBSIDY AND VALUE INCREASES.

Source of Variation	Sum of Squares	Degrees of Freedom	Mean Square	F
Regression	13.6425	1	13.6425	16.914
Residual	6.4525	$n - 2 = 8$	.8066	
Total	20.1	$n - 1 = 9$		

conclusion that the researcher can draw from Figure 12.5 is that there is 99 percent confidence that the regression line is a true improvement over the mean and potentially useful for predictive purposes subject to the discussion that follows.

Coefficient of Determination (r^2)

Some researchers stop at this point and use regression lines for predictive purposes when and if the line proves to be significant. Significance, they argue, indicates that the line is a better representation of the relationship between the independent variable and the dependent variable than is the mean of the dependent variable that bears no relationship to the independent variable.

A crucial weakness with such an approach is that significance indicates only that a relationship exists—not how strong it is. It is frequently the case that a regression line reflects only a very weak relationship that is significant in that it is an accurate reflection of the relationship, but the mere existence of such a relationship may not merit its use as a tool for making predictions and projections.

The question then becomes, "If mere significance is not sufficient to justify projection using regression analysis, what measure should be used to make that determination?"

Just as analysis of variance has a correlation ratio (E^2) that indicates the amount of the total variance in the dependent variable explained by the categories of the independent variable, regression analysis uses a coefficient of determination (r^2) to indicate the amount of the total variance that the regression line explains. The coefficient of determination is calculated as follows:

$$r^2 = SS_{regression} / SS_{total} \qquad (12.12)$$

For the subsidy/value example,

$$r^2 = 13.6425/20.1$$

$$= .679$$

Therefore, approximately 68 percent of all the variance in market value increases, in the general population to which the sample refers, is explained by the amount of rehabilitation subsidies that flow to the properties and the blocks on which they are located. To expand further and more specifically on the question above, "Is a 68 percent explanatory rate sufficient to justify use of the regression line for making predictions?"

For this latter question, there is no generally accepted statistical principle or convention as there are with 95 percent and 99 percent confidence levels, for example. Some of the more conservative statisticians favor using a 95 percent explanatory power (r^2) before they would allow prediction; however, they also agree that 95 percent explanation is rarely achieved, thereby rendering regression analysis informative as to the nature of the relationship between interval variables but useless as a predictive tool. We, in an effort to make use of this valuable tool but not when its powers are substantially weakened, recommend the use of regression analysis for prediction when the regression line is statistically significant and when its coefficient of determination is at least equal to 0.5, or 50 percent.

Multiple Regression

Rarely is one independent variable sufficient to achieve an r^2 of .5, and even when it does (as in the case above of the subsidies and value increases), why stop there? Why not add more independent variables to the analysis in order to achieve even greater explanatory power? This is indeed possible. For example, assume that a researcher, having achieved an r^2 of .68, now seeks an even better model of value increases and hypothesizes that overall citywide value increases might also be at play, as might percentage of ethnic minority residents in the census tract. The researcher finds the information set out in Figure 12.6.

Manual calculation of a new regression line, if the new variables prove to be significant, is much too burdensome for this book's readers and most researchers. Computers, thankfully, carry this load for us now. The computer determines which among the three variables is the greatest contributor to explaining value increases (highest r^2). The computer then adds the next contributor, and so forth, until there are no significant variables remaining. In the case above, minority percentage of the census tract proved to be the most important contributor ($r^2 = .707$), and then citywide value increases added another .134 beyond minority percentage for a total $R^2 = .841$. Note the change from r^2 to R^2: r^2 is the *coefficient of determination* for one independent variable, and R^2 is the *multiple coefficient of determination* when the regression line contains more than one independent variable. Subsidies are no longer significant. That is, citywide value increases and ethnic percentages are the real explainers of specific block-level increases, and the new regression line (as determined by computer) is $y = 3.523 - 5.288x_1$ (minority proportion) + $.246x_2$ (citywide value increases in thousands). Therefore, the property that is represented by the sample can be expected to increase in value by approximately $250 for every $1,000 in citywide increases and decrease in value by $50 for every 1 percent increase in minority percentage within its census tract.

FIGURE 12.6. REHABILITATION SUBSIDY/MARKET VALUE INCREASE ANALYSIS.

Block	Rehabilitation Subsidy (in thousands)	Mean Market Value Increase (in thousands)	Citywide Value Increases (in thousands)	Percentage Ethnic Minority Residents
A	100	3	3	.28
B	85	3	6	.35
C	95	1	3	.51
D	130	5	8	.10
E	110	5	10	.25
F	75	2	5	.68
G	135	4	9	.46
H	60	0	5	.71
I	120	4	10	.20
J	115	3	4	.38

The subsidy variable, no longer a part of the regression line, has demonstrated no statistical value beyond these other variables once they are introduced and can be considered to be highly correlated with citywide increases and ethnicity, explaining the same components of the dependent variable. By itself, subsidies explain 68 percent of the variance in value increases; however, when more variables are introduced, it is found that minority percentage explains almost 71 percent of these block-level increases. With subsidies being eliminated as a significant variables, it is possible that cities with higher minority percentages may have had less money available for rehabilitation subsidies, thereby explaining the high correlation between the two.

Furthermore, by itself, citywide value increases explain almost 71 percent of the variance , but after accounting for its correlation with ethnic minority percentage, it added only 13.4 percent beyond the explanatory power of minority percentage.

Multiple regression is but one of many tools in more advanced statistics for use in multivariate analysis. It is a fascinating part of statistics and one that we encourage readers to pursue. These more advanced statistics are, however, beyond the scope of this more fundamental survey research book.

Dummy Variables

Regression analysis is designed only for interval variables, and yet there are variables that are not interval and might be able to contribute significantly to a regression model. For example, if individual income data are being examined for purposes of predicting earnings, gender or ethnicity might be key predictors, but they are not interval. They can, however, be reconfigured for regression purposes as binary variables, coded as 0 or 1 only, with the 0 representing the absence of a characteristic and a 1 representing its presence.

In order to accomplish this, a variable called a *dummy variable* (sometimes referred to as *indicator variables*) is created for $k - 1$ categories of the independent variable. If gender is the variable, then the two categories ($k = 2$)—male and female—can be accommodated by creating one new dummy variable ($k - 1 = 1$), where a code of 1 would indicate female and 0 would indicate male. If ethnicity is the issue and there are considered to be five ethnic categories (white, African American, Latino, Asian, other), then four dummy variables would be required, as follows:

If $x_1 = 1$, and x_2, x_3, and $x_4 = 0$, then the individual is white.

If $x_2 = 1$, and x_1, x_3, and $x_4 = 0$, then the individual is Latino.

If $x_3 = 1$, and x_1, x_2, and $x_4 = 0$, then the individual is African-American.

If $x_4 = 1$, and x_1, x_2, and $x_3 = 0$, then the individual is Asian.

If x_1, x_2, x_3, and $x_4 = 0$, then the individual is of another ethnicity.

Let it be assumed that multiple regression analysis produces the following line:

$$y = 30{,}000 + 5{,}000x_1 - 3500x_3 - 7{,}000x_5 \text{ (female)} + 1{,}000x_6 \text{ (years of education)}$$

This line can be interpreted as follows. The intercept, $30,000, is the base income, to which $5,000 is added for predictive purposes if the individual is white or $3,500 is deducted if the individual is African American, $7,000 is deducted for a female, and $1,000 is added for every year of education. Being Latino, Asian, or of another ethnic group is not significant in this example. A researcher would therefore predict that a white female with sixteen years of education would earn $44,000 and that an African American male with fourteen years of education would earn $40,500.

Dummy variables also serve very well as what are called *intercept variables*, where a significant event has disrupted the trend that regression relies on. For instance, if in the year 2010, Thailand is analyzing its attractiveness to tourists, the tragic tsunami at the end of 2004 will have obviously had an impact on it for some period thereafter. A dummy variable, where $0 = 2004$ and before and $1 = 2005$ and thereafter, would work well to account for this terrible event.

Pearson's *r* Correlation

The strength of the relationship between two interval scale variables can be measured by *Pearson's r*, or what is commonly referred to as the *coefficient of correlation*. Pearson's *r* can be calculated in accordance with the following formula:

$$r = \frac{\sum (x - \bar{x})(y - \bar{y})}{\sqrt{(x - \bar{x})^2 (y - \bar{y})^2}} \tag{12.13}$$

The calculation of Pearson's *r* for the data in Figure 12.1, based on equation 12.13, is demonstrated below for the original rehabilitation subsidy/property value example. It should be noted that Worksheet 12.1, which was used to calculate the regression line, provides all of the information that is necessary to make this calculation, in which the numerator is the sum of the products (*SP*) and the denominator is the square root of SS_x multiplied by SS_y.

$$SP = 267.5$$
$$SS_x = 5,262.5$$
$$SS_y = 20.1$$

$$r = 267.5 / \sqrt{(5,262.5)(20.1)}$$
$$= 267.5 / \sqrt{105,776.25}$$
$$= 267.5 / 325.233$$
$$= .82$$

A Pearson's *r* of .82 can be interpreted in accordance with the guidelines presented in Figure 10.3 (interpretation of calculated gamma) as a very strong positive association. In other words, rehabilitation subsidies are strongly correlated positively with property value increases: the more money that is available, the greater is the increase in property values.

It is also possible to determine Pearson's *r* from the coefficient of determination: $-r^2$. Pearson's *r* is, as would be supposed, the square root of r^2 with a sign (positive or negative) that matches the slope of the line. The r^2 from the simple linear regression of subsidies and value increases was determined above to be .679, the square root of which is .82, and the sign is positive to match the positive $b = .051$.

As with any other calculation based on a sample, statistical significance of Pearson's r must be established. If Pearson's r is derived as a by-product of regression analysis, the significance of the regression line will also indicate whether Pearson's r is significant.

If Pearson's r is determined by itself, as would be the case when the nature and strength of the relationship between two interval variables is desired and a line for predictive use is not needed, then its significance can be tested using a t test, as follows:

$$t = (r\sqrt{n-2}) \div \sqrt{(1-r^2)} \qquad (12.14)$$

$$df = n - 2$$

In the subsidy/value example:

$$t = .82(\sqrt{8}) \div \sqrt{(1 - .82^2)}$$
$$= 2.319 \div \sqrt{(1 - .6724)}$$
$$= 2.319 \div \sqrt{3,276}$$
$$= 2.319 - .5724$$
$$= 4.05$$
$$df = 8$$

The critical t (from Exhibit 7.2) at 95 percent and 99 percent—two tail because t can be positive or negative—is, respectively, 2.306 (95 percent) and 3.355 (99 percent), and significance is established by the calculated $t = 4.05$, as was the case when regression analysis was performed.

Choosing Among Tests of Significance and Measures of Association

The significance tests and measures of association discussed in this book have been presented in the context of their application to research questions involving variables of various levels of measurement. There have been many such tests, and we believe the summary in Exhibit 12.1 will be useful.

EXHIBIT 12.1.   CHOOSING THE APPROPRIATE
STATISTICAL TEST OR MEASURE.

Level of Measurement

Independent Variable	Dependent Variable	Notes and Comments	Test of Significance	Measure of Association
Nominal	Nominal		Chi-square	Cramér's *V*
Nominal	Ordinal		Chi-square	Cramer's *V*
Nominal	Interval		Analysis of variance or independent samples/paired samples *t* tests	Eta
Ordinal	Nominal		Chi-square	Cramér's *V*
Ordinal	Ordinal	If analyzing trend	Gamma	Gamma
Ordinal	Ordinal	If not analyzing trend	Chi-square	Gamma
Ordinal	Interval		Analysis of variance or independent samples/paired samples *t* tests	Eta
Interval	Nominal		Chi-square	Cramér's V
Interval	Ordinal	If analyzing trend	Gamma	Gamma
		If not analyzing trend	Chi-square	Gamma
Interval	Interval	If prediction	Regression analysis	Pearson's *r*
Interval	Interval	If no prediction	Analysis of variance or independent samples/Paired samples *t* tests	Pearson's *r*

EXERCISES

1. A governmental analyst is trying to identify the relationship, if any, between the level of poverty in a city and the number of gang-related offenses. The analyst has chosen 10 cities throughout the state and has found the following information:

City	Gang-Related Offenses	Poverty Rate
A	1,000	4.2
B	1,200	5.0
C	800	4.3
D	1,300	3.9
E	2,000	5.1
F	3,000	8.7
G	1,100	3.5
H	200	4.0
I	2,700	6.1
J	500	5.0

    a.   What is the research hypothesis you would suggest to the analyst?
    b.   Determine the regression line for these data.
    c.   Is the regression line statistically significant?
    d.   What is the importance of a statistically significant regression line?
    e.   Predict the number of gang-related offenses in a city with a poverty rate of 8.0 percent.
    f.   Calculate the coefficient of determination and explain.

2.   A planning analyst is trying to identify the relationship, if any, between the level of pollutants in the air of a city and the number of tourists who visit. Her theory is that the more significant the tourist industry, the less reliant that city needs to be on smokestack industries, and therefore the less pollution. The analyst has chosen 10 cities with populations between 250,000 and 500,000 throughout the state and has found the following information:

City	Pollutant Count	Tourists per Fixed Period
A	30	100
B	12	200
C	11	225
D	5	400
E	27	200
F	10	350
G	2	500
H	8	450
I	20	150
J	13	375

    a.   Determine the regression line for these data.
    b.   Is the regression line statistically significant?
    c.   What is the importance of a statistically significant regression line?
    d.   Predict the pollutant count for a city with 275 tourists.
    e.   Calculate the coefficient of determination and explain.

3.   The Economic Development Department of the City of Good Business Climate, U.S.A., has undertaken a development strategy whereby it is offering outright subsidies to local businesses in an effort to keep them from moving

to their rival city, We'll Give You Anything, U.S.A. The new city council, however, has called this practice into question and has asked that you compile a list of such subsidies over the past 10 years and compare that list to the city's unemployment rate. The council wants to know if there has been a positive impact by these subsidies on the unemployment rate and what the city can expect if it continues the policy.

Your research has found the following information:

Year	Unemployment Rate	Prior Year Subsidy
1994	4.2	$2.7 million
1995	3.7	3.2 million
1996	3.5	3.8 million
1997	3.9	4.1 million
1998	4.1	4.0 million
1999	5.0	3.0 million
2000	4.5	3.2 million
2001	4.8	3.3 million
2002	5.5	3.0 million
2003	5.8	2.7 million

a. Assuming that your data are statistically significant, what would you predict the 2005 unemployment rate to be if the total subsidy budget for 2004 was $2.9 million?

b. Is the regression line statistically significant?

c. How much of the total variance in your dependent variable does the independent variable explain?

d. What is the value of the Pearson's r for these data?

4. A planning analyst is trying to identify the relationship, if any, between applications for land use permits per week and the annual inflation rate. The analyst has chosen 10 years' worth of data for the study and has found the following information:

Year	Land Use Permit Applications	Inflation Rate
1995	40	5.2
1996	30	4.0
1997	25	5.3
1998	50	2.9
1999	45	6.1
2000	10	7.7
2001	30	5.0
2002	40	2.8
2003	100	4.5
2004	25	4.2

a. Calculate Pearson's correlation coefficient for these data, and describe the strength and direction of the relationship.

b. Is the association statistically significant?

For Exercises 5 to 7, indicate the regression line of best fit, determine whether that line is statistically significant, and calculate and interpret the coefficient of determination.

5. Ten families, as follows:

Number of Children	Gallons of Milk Consumed
2	2
3	4
1	1
0	1
2	5
4	7
1	2
0	0
5	5
2	3

Predict the milk consumed by a family of 6 children.

6. Same families indicate the following educational attainment by the most educated adult in the household:

Years of Education

12
10
14
17
14
12
16
18
8
9

Predict the number of children in a household where the most educated adult has 15 years of education.

7. Yogi Berra once said, "Baseball is 90 percent pitching and the other half is luck." Let's see if he's right. Below are the 1993 earned run averages (ERA) for the pitching staffs of the 14 National Baseball League teams and also their team batting averages for that year. Also below are the number of wins each team had in that year. Indicate through regression analysis whether Yogi was right. Make certain to perform all of the calculations indicated above for both ERA and batting average. Predict the number of wins for a team with a staff ERA of 3.90. Also predict the number of wins for a team with a batting average of .280 (See the note below on batting average.)

Team	ERA	Batting Average*	Wins
Los Angeles	3.50	61	81
San Diego	4.23	52	61
San Francisco	3.61	76	103
Colorado	5.41	73	67
Atlanta	3.14	62	104
New York	4.05	48	59
Philadelphia	3.95	74	97
Florida	4.13	48	64
Montreal	3.55	57	94
Chicago	4.18	70	84
Cincinnati	4.51	64	73
Pittsburgh	4.77	67	75
Houston	3.49	67	85
St. Louis	4.09	72	87

*Batting averages are generally given as three-digit proportions, such as .254 Rounding errors for such data, when squared can be substantial. Therefore, the averages given are "points" above .200 and are less subject to such errors. For instance, San Diego's batting average of 52 equates to an actual average of .252, Los Angeles's to .261, and so on.

8. You have a friend who has come up with a strange theory. He says that the number of barks made by a dog each day are predictable from the number of times each month that the dog is fed table scraps: the more scraps, the more barks (they get spoiled and insistent on more). You decide to check out this theory by studying your neighbors' dogs. After all, there are serious city planning implications in this analysis in terms of noise issues. You record the behavior of the following dogs in your area and find the following:

Dog	Number of Barks Each Day	Table Scrap Feedings per Month
Laddie	77	56
Run Ton Ton	36	42
Scoopy Dew	86	49
Neil	132	77
Bob	22	35
Huckleberry	15	35
Tramp	49	14
Snoop	183	70
Deputy Dawg	58	49
Old Screamer	1	28

a. Assuming that your data are statistically significant, what would you predict the number of daily barks to be for a dog who gets table scraps 50 times per month?
b. Is your predictive equation statistically significant?

c. How much of the total variance in your dependent variable does the independent variable explain?

d. What is the value of the Pearson's *r* for these data?

e. "This is absolutely ridiculous," says a voice from behind you. "You call yourself a city planner? Everybody knows that dogs bark only when they are frightened. It has nothing to do with food, and nothing frightens dogs more than silence. That's why they bark: to keep themselves company and to chase away the fear of silence. You can measure the silence in an area by the absence of construction going on, and no construction takes place on days when building inspectors inspect. *The more inspectors in the neighborhood on a given day, the more barks. It's just that simple.*"

The critic pulls out his study, which shows a simple linear regression coefficient of determination of .57. Do we have enough information to identify which has a stronger explanatory power: table scraps or building inspectors? If so, which seems to be the better explanation of dog barks? Explain your answer.

CHAPTER THIRTEEN

PREPARING AN EFFECTIVE FINAL REPORT

The final report is the vehicle for communicating to the audience the conclusions and recommendations derived from the study. It should be viewed by the researcher as integral to the survey research process as a whole; therefore, the research process should not be considered complete until the final report has been prepared and disseminated. Within the report, the analysis of the data, including tables, graphs, and other statistical presentations, should be well organized and clearly explained so that the intended audience can comprehend the essential findings of the study. This chapter suggests an appropriate format and useful guidelines for preparing a formal report of survey research findings.

Report Format

Several fundamental considerations must be taken into account as the final report is prepared.

The Title

The report should have a clear and succinct title that identifies the focus of the research. Frequently a subtitle can help clarify the subject matter of the research further. For example, a research report concerning growth control limitations in

EXHIBIT 13.1.   EXAMPLE OF A COVER PAGE.

Issues Concerning Economic Growth:
A Critique of Oswego County's Proposed
Countywide Impact Fee Program

Prepared by
XYZ Research Corporation
for the
Oswego County Taxpayers Association

March 2005

New York State was titled, "Issues Concerning Economic Growth: A Critique of Oswego County's Proposed Countywide Impact Fee Program." The title should appear on the cover of the report, along with the names and affiliations of the authors. The date of report dissemination should also be included, as should the client or sponsor for whom the study was conducted. Exhibit 13.1 is an example of a well-constructed cover page. The title should also appear within the report, either on a separate title page or at the top of the first page of the report.

Executive Summary

The reader of the report frequently finds it helpful when a short summary of findings is included at the beginning of the report. This summary can serve as a source of reference after the report has been read, and it can also serve as the basic source of information about the report itself for audiences that are interested in only the main findings and conclusions, not the details of the subject matter. The executive summary can be as short as one page or as long as ten pages, depending on the length and complexity of the report. We have found that three to five pages is the desirable length. The summary should very briefly introduce the report and indicate the methodology employed in the survey but move ahead quickly to outline key findings that are divided into appropriate subsections and presented in a bulleted format. Graphs and charts are generally not included in the summary.

Introduction to the Study

The report should start by providing the audience with some background about the subject matter of the study and by placing the study in appropriate perspective with regard to the history and current significance of the research topic. Major social and political events of the time that the researcher feels have some bearing on the responses should be presented in the introduction. The introduction must contain a clear statement of the specific purpose of the study, including a description of the issues and an explanation of why the researchers decided to pursue the subject in this fashion and at this time.

Review of Preliminary Research

The initial processes undertaken by the researcher in identifying the research focus and helping to develop the actual research instrument should be summarized. This includes a review of existing literature consulted and a discussion of the key groups and individuals who participated in the development of the information base from which the pretest and draft questionnaire evolved. The specific questions at issue or research hypotheses to be tested should be stated and shown as having been directly derived from this preliminary research. The hypotheses should be presented in the context of the information required from the study. That is, the researcher should indicate what the research is designed to discover that was heretofore unknown.

Method of Research

It is important to include an explanation of the methodology that was employed to obtain and analyze the data. For sample survey research, there are three methodological issues: sample selection, survey procedure, and data analysis.

Sample Selection. The report should detail the procedures employed in selecting the sample. This discussion must include an explanation of how an appropriate working population was identified to represent the general population. The determination of this sampling frame should be accompanied by a discussion of any potential systematic biases. Also to be included in this discussion are the determination of sample size (specifying the level of confidence and margin of error) and the specific sampling method employed in the selection of the final sample.

Survey Procedure. The survey method should be discussed, including the recruitment and selection of interviewers for telephone and in-person surveys, procedures employed in the initial mailing for mail-out surveys, the Web format and method of dissemination employed, follow-up procedures, response rates, and the time frame of the study. Patterns of biases that may have been identified in the interviewing process must be identified and explained. Their potential effect on survey results should be indicated.

Data Analysis. The researcher should briefly describe the statistical methods used in data analysis, including all applicable tests of significance and measures of association. Included in this section are explanations of the meaning and importance of these tests and measures. The margin of error should be mentioned once again in this section, as should the level of confidence, which is generally 95 percent.

Survey Research Findings

The major part of the report consists of the research findings. This portion is composed largely of tables and graphs, with appropriate descriptive and analytical statistics accompanied by written explanations of the tabular and graphic results. A table or graph should appear in the text as close to its initial mention as possible. It has been found that this placement of tables and graphs within the text lends itself to increased convenience for readers, who can study the researcher's interpretation of the data while maintaining ready visual access to the data themselves. All charts and tables should be numbered to allow for cross-references within the text of the report.

Charts or tables should be prepared for each survey question, and contingency tables should be presented at the researcher's discretion within the framework of the research issues under consideration. This discretion should be guided by the relevance of information, the statistical significance of any apparent relationships among the variables involved, and the relative strength of the variables' association.

Focus Group Findings

When focus groups are used as part of the research process, the findings from these discussions must be incorporated into the final report. This portion of the report should include a description of the number of sessions, the key characteristics of the participants, and the date, time, and place of each session. The structured question format used for the sessions should be presented, and the notes taken at each focus group session, refined by reviewing the videotape of the session, should be recorded in the form of summary minutes. These minutes constitute a listing, by participant or subject matter, of the key remarks offered in the session. Finally, a composite summary of the findings from all the sessions should be provided, with common themes particularly highlighted.

Conclusions

The report should conclude with a strong section that draws implications from the findings, indicates relationships and trends among the various tables and graphs, and relates the findings from the study to any relevant previous studies or literature. When appropriate, policy recommendations should be put forth, and opportunities for further research should be discussed.

Bibliography

Important publications and documents consulted during the research process should be listed in a bibliography at the end of the report.

Appendixes

Certain information should be attached to the report in the form of an appendix. Such material always includes a copy of the survey instrument itself and a full set of frequency distributions. Other potential appendixes may be a list of verbatim open-ended responses and detailed explanations of certain statistical techniques or sampling procedures, including all applicable mathematical equations (which are much better placed in the appendix than in the body of the report itself).

There is a considerable amount of discretion involved in choosing the material to include in an appendix. The overriding principle is to provide material that the researcher feels would be beneficial to the reader but would tend to compromise the readability of the report. This allows the researcher to maintain a pleasant communicative flow in the report itself while still providing all the important information to readers.

Specific Considerations for Formally Reporting Survey Results

Within the report framework, there are further issues, particularly with regard to format and style of writing, that the researcher should address in the preparation of the final report. These issues can be organized into two general categories: (1) vocabulary, jargon, and statistical notation and (2) reporting of numerical detail.

Vocabulary, Jargon, and Statistical Notation

The audience or client must be considered in deciding the writing style and the extent of professional vocabulary to be used. If a report is prepared for a technically oriented audience (for instance, engineers, doctors, scientists, or accountants), the researcher may feel more comfortable using terms that are regarded as specific to their particular discipline. The more general or diversified the audience, however, the less technical the language should be, as long as meaning and substance are not sacrificed. An example of such a general audience would be the people who would read a research report summarizing the results of a community public opinion survey.

The report should not rely on statistics and statistical notation as substitutes for descriptions of the relationships involved. For example, it is more informative to an audience to say that there is a moderate association between the ethnicity and income of the citizens in a community than to write only that the ethnicity/income Cramér's $V = .35$. In general, it is better to use descriptive words (for instance, *mean, test of significance,* and *measure of association* or *correlation*) than to rely entirely on statistical notations such as $\bar{x}$, χ^2, or V. Statistical notations can be used in the written portion of the report, however, but should follow immediately after the technique's mention. For instance, when chi square is mentioned, it can be followed by the parenthetical reference (χ^2), as indicated in the following example: "The chi-square test of significance (χ^2) for the relationship between political party affiliation and sex of the respondent did not establish the existence of a statistically significant relationship between the two variables."

Word choice is very important. Be careful not to give the impression that survey findings are universal. Instead of writing "The people feel that . . . . ," write, "Most people (76 percent) feel that . . . . ," because it is highly unlikely that all people have the same opinion. Be careful when using words such as *strong* or *significant;* these words and certain others connote statistical relationships that may mislead the reader. It is less confusing to indicate that a particular issue has "substantial support" rather than "significant support" when reporting results. This is especially true in reporting frequency distributions. The researcher must also be careful to avoid the use of emotive or judgmental words—words that denote surprise, discomfort, or displeasure. For example, a report should not contain the phrase, "It was particularly upsetting to find that . . . ."

For ease of reading, footnotes should be used sparingly. Important material should be incorporated into the body of the text as much as possible. Reference citations should always be embodied within the text in accordance with the (author, year, page) format. Each citation should correspond to a full bibliographical reference at the end of the report. Content notes (explanatory digressions that, in the judgment of the researcher, would tend to obstruct the flow of the text) should also be used as infrequently as possible, but when they are used, they should be placed at the bottom of the page as a footnote or at the end of the chapter as an endnote for ready reference.

Reporting of Numerical Detail

The written portion of the report should indicate percentages, fractions, or ratios rather than absolute frequencies. That is, rather than reporting that 475 Democrats oppose gun control, it is more informative to say that approximately 66 percent, or nearly two-thirds, of the Democrats surveyed oppose gun control. When the overall sample is quite small or a particular subgroup within the sample is small, the researcher should report both the percentages and the corresponding frequencies. For example, rather than reporting that "40 percent of African American office workers favor a change in their union contract," it is more appropriate to report that "in a survey of twenty office workers, ten of whom are African American, four African American workers (40 percent) favor a change in their union contract." In this way, the reader will not be misled into believing that this particular finding is more substantial than it actually is.

Use whole numbers and common fractions whenever possible. Instead of reporting that "men favor an issue more than women by a ratio of 1.85 to 1," it is better to say that "nearly twice as many men as women favor this issue (1.85:1)." This approach lends itself to significantly more pleasant reading, while simultaneously providing the audience with the specific numerical detail. Similarly, it is better to report that approximately one-fourth of a population expressed a certain attitude than that seven thirty-seconds did (which may be more accurate but is much more difficult to translate into a commonly understood quantity).

Research reports should predominantly use charts and tables to report results, with charts used more frequently and tables reserved for more complex findings. Charts and graphs will cause the reader to recognize the findings much more easily and in a way that is more pleasing. Color and three-dimensional effects tend to promote this advantage even more. For example, Figure 13.1 shows that almost three-fourths of these respondents favor, at least somewhat, the building of a seawater desalination facility. Figure 13.2 indicates ratings of police response to emergency and nonemergency calls. Figure 13.2 makes use of a five-point scale from excellent to very poor (where 1 = excellent and 5 = very poor) and also indicates mean ratings based on that scale. It can easily be seen that respondents rate emergency calls higher than nonemergency calls. This is derived from means of 2.0 and 2.6, respectively, and from 75 percent good or better ratings for emergency calls versus 54 percent for nonemergency calls.

FIGURE 13.1.   FAVOR OR OPPOSE BUILDING SEAWATER DESALINATION FACILITIES.

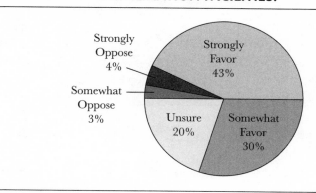

FIGURE 13.2.   RATING OF POLICE RESPONSE TO EMERGENCY AND NONEMERGENCY CALLS.

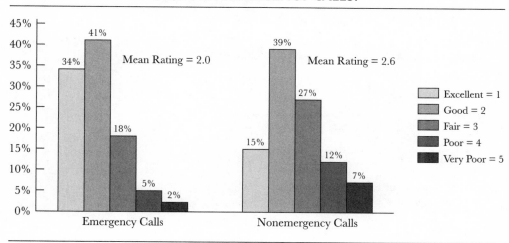

Charts and graphs can be more complex, such as Figure 13.3, which depicts mean satisfaction ratings for various characteristics of bus service. The figure includes mean ratings on a 1 to 5 scale and percentages that indicate the higher trip ratings for each characteristic. It can be seen with relative ease that route convenience and safety are more satisfactory to riders than are waiting time and the timeliness of the buses.

FIGURE 13.3. MEAN SATISFACTION RATINGS FOR VARIOUS FEATURES OF BUS SERVICE.

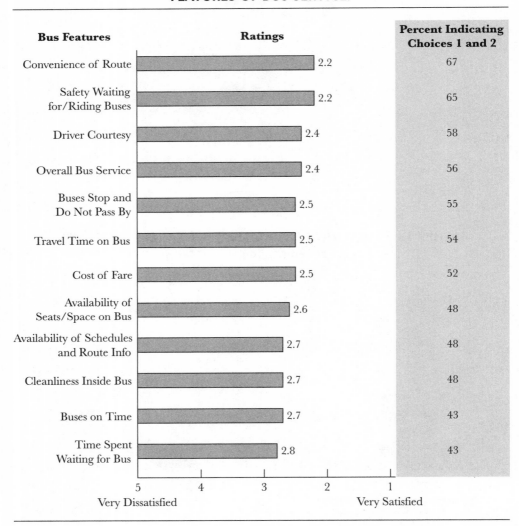

Bus Features	Ratings	Percent Indicating Choices 1 and 2
Convenience of Route	2.2	67
Safety Waiting for/Riding Buses	2.2	65
Driver Courtesy	2.4	58
Overall Bus Service	2.4	56
Buses Stop and Do Not Pass By	2.5	55
Travel Time on Bus	2.5	54
Cost of Fare	2.5	52
Availability of Seats/Space on Bus	2.6	48
Availability of Schedules and Route Info	2.7	48
Cleanliness Inside Bus	2.7	48
Buses on Time	2.7	43
Time Spent Waiting for Bus	2.8	43

5   4   3   2   1
Very Dissatisfied   Very Satisfied

Another more complex feature of graphical depictions is the *call out,* which provides more detailed information about a specific aspect of the chart or graph. Figure 13.4 shows (in the stacked column in the lower right) that among the 61 percent of all households that have automatically controlled sprinkler systems for their lawns (as shown in the pie chart), 43 percent of the 61 percent adjust their sprinklers 4 or more times per year.

FIGURE 13.4.   HOUSEHOLD HAS AUTOMATICALLY CONTROLLED SPRINKLER SYSTEM FOR LAWN.

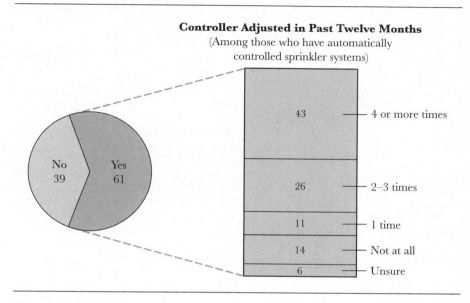

Controller Adjusted in Past Twelve Months
(Among those who have automatically
controlled sprinkler systems)

No 39

Yes 61

43 — 4 or more times

26 — 2–3 times

11 — 1 time

14 — Not at all

6 — Unsure

Sometimes the information to be conveyed is too intricate or voluminous for portrayal in a chart or graph. Such situations require tables. For instance, Table 13.1 shows potential uses for recycled water, the overall mean rating for each such use, the groups that favor that particular use of recycled water more (including their mean ratings), and the groups that are more opposed (including their means). Table 13.1 indicates that use of recycled water is most favored for freeway landscaping and golf courses, with higher-income, better-educated, and white respondents more in favor than respondents of lower income, less education, and Latinos.

Once again, it is important to state that all charts and tables that are included in the report should be numbered and titled and receive some comment in the text. The researcher should avoid reporting an excessive amount of detail and should instead selectively report the few salient details that bear most directly on the focus of the study. Rather than reporting detail as follows, "Concerning mode of transportation to work, 68.2 percent use automobiles, 23.7 percent use public transit, 4.8 percent walk, 2.9 percent ride bicycles, and 0.4 percent use taxis and other dial-a-ride services," the researcher should report the information by stating that "over two-thirds of the population (68 percent) use automobiles, and nearly one-fourth (24 percent) use public transit." If the research has a particular focus regarding one

TABLE 13.1. STATISTICALLY SIGNIFICANT DIFFERENCES FOR VARIOUS POTENTIAL USES OF RECYCLED WATER (SCALE: 1 = STRONGLY FAVOR TO 5 = STRONGLY OPPOSE).

Potential Uses of Recycled Water	Overall Mean Index	Statistically Significant Differences			
		Stronger Support/ Weaker Opposition		Weaker Support/ Stronger Opposition	
Freeway/golf landscape	1.35	Whites	1.25	Hispanics/Latinos	1.61
				Mixed ethnicities	1.67
		Income $100K+	1.18	Income under $25K	1.63
		Some college/more	1.29	High school/less	1.66
Toilets in new buildings	1.57	Ethnicities other than Hispanic/Latino and mixed ethnicity	1.48	Hispanics/Latinos	1.88
				Mixed ethnicities	1.87
				High school diploma	
		Some college or more	1.49	or less	1.92
Electronics manufacturing	1.66	Some college/more	1.60	High school/less	1.90
Industrial manufacturing processes	1.69	Ages 45 or more	1.52	Ages 18–44	1.88
		Whites	1.49	Blacks/African Americans	2.00
		Asians	1.59	Hispanics/Latinos	2.08
		Some college/more	1.61	High school/less	2.12
Sports fields/ parks	1.70	Asians	1.43	Hispanics/Latinos	2.04
		Whites	1.60	Blacks/African Americans	2.27
		Males	1.56	Females	1.84
Multifamily common areas	1.76	1–2 person households	1.56	3+ person households	1.97
		Whites	1.55	Hispanics/Latinos	2.49
		Asians	1.59	Blacks/African Americans	2.35
		Own	1.65	Rent	2.01
		Males	1.63	Females	1.90
Residential front yards	1.92	Asians	1.73	Hispanics/Latinos	2.38
		Whites	1.60		
Agricultural irrigation	2.15	Whites	1.96	Hispanics/Latinos	2.86
		Asians	1.78		
		Males	1.95	Females	2.36

of the other modes of transportation, it should be commented on, but otherwise, the written portion of the report should highlight only the critical findings.

EXERCISES

1. Figure 13.5 depicts opinions about whether desalination will harm the ocean. Using this figure, write a paragraph for inclusion in a final report that summarizes and highlights these opinions. Be sure to follow the guidelines outlined in this chapter.

FIGURE 13.5.   BELIEVE SEAWATER DESALINATION PROCESS IS HARMFUL OR NOT HARMFUL TO THE LOCAL OCEAN ENVIRONMENT.

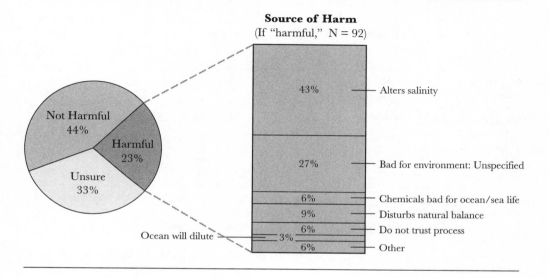

TABLE 13.2.   AREA WHERE CONVENIENCE GOODS ARE PURCHASED MOST FREQUENTLY BY ETHNIC GROUP.

| | Ethnic Group | | | | | | | | | |
| | White | | Black | | Asian | | Hispanic | | Total | |
Commercial Area	f	%	f	%	f	%	f	%	f	%
City Heights	195	44.1	65	51.6	18	47.3	5	4.5	283	39.5
Midcity (other than City Heights)	119	26.9	30	23.8	10	26.3	27	24.3	186	25.9
Nearby regional shopping center areas[a]	60	13.6	17	13.5	5	13.2	16	14.4	98	13.7
Other[b]	68	15.4	14	11.1	5	13.2	63	56.8	150	20.9
Total	442	100.0	126	100.0	38	100.0	111	100.0	717	100.0

Note: $\chi^2 = 3.80$, not significant at 95 percent (critical $\chi^2 = 16.92$, $df = 9$). Convenience goods include groceries, medical supplies, do-it-yourself products, and dining out.

[a]Nearby shopping centers include Mission Valley (Mission Valley), Fashion Valley (Mission Valley), Horton Plaza (Downtown), and Grossmont (La Mesa).

[b]"Other" responses include various locations that are generally closely tied to the location of the respondent's workplace.

TABLE 13.3. AREA WHERE SHOPPING GOODS ARE PURCHASED MOST FREQUENTLY BY ETHNIC GROUP.

| | Ethnic Group | | | | | | | | | |
| | White | | Black | | Asian | | Hispanic | | Total | |
Commercial Area	f	%	f	%	f	%	f	%	f	%
City Heights	28	6.6	20	16.0	8	20.5	11	9.9	67	9.6
Midcity (other than City Heights)	115	27.1	37	29.6	9	23.1	27	24.3	188	26.9
Nearby regional shopping center areas[a]	175	41.3	44	35.2	11	28.2	51	46.0	281	40.2
Other[b]	106	25.0	24	19.2	11	28.2	22	19.8	163	23.3
Total	424	100.0	125	100.0	39	100.0	111	100.0	699	100.0

Note: $\chi^2 = 19.86$, significant at 95 percent (critical $\chi^2 = 16.92$, $df = 9$). Shopping goods include appliances, specialty goods, clothing, furniture, toys, and sporting goods.

[a]Nearby shopping centers include Mission Valley (Mission Valley), Fashion Valley (Mission Valley), Horton Plaza (Downtown), and Grossmont (La Mesa).

[b]"Other" responses include various locations that are generally closely tied to the location of the respondent's workplace.

2. Tables 13.2 and 13.3 examine shopping patterns among various ethnic groups residing in the community of City Heights. The residents were asked where they shopped most often for convenience goods, which include groceries, medical supplies, and hardware (Table 13.2), and where they shopped for shopping goods, which include appliances, furniture, and clothing (Table 13.3). Write an account, for inclusion in a final report, that addresses the most salient relationships in these tables.

TABLE OF AREAS OF A STANDARD NORMAL DISTRIBUTION

(A) Z	(B) Proportion of Area Between Mean and Z	(A) Z	(B) Proportion of Area Between Mean and Z
0.00	0.0000	0.20	0.0793
0.01	0.0040	0.21	0.0832
0.02	0.0080	0.22	0.0871
0.03	0.0120	0.23	0.0910
0.04	0.0160	0.24	0.0948
0.05	0.0199	0.25	0.0987
0.06	0.0239	0.26	0.1026
0.07	0.0279	0.27	0.1064
0.08	0.0319	0.28	0.1103
0.09	0.0359	0.29	0.1141
0.10	0.0398	0.30	0.1179
0.11	0.0438	0.31	0.1217
0.12	0.0478	0.32	0.1255
0.13	0.0517	0.33	0.1293
0.14	0.0557	0.34	0.1331
0.15	0.0596	0.35	0.1368
0.16	0.0636	0.36	0.1406
0.17	0.0675	0.37	0.1443
0.18	0.0714	0.38	0.1480
0.19	0.0753	0.39	0.1517

(A)	(B)	(A)	(B)
Z	Proportion of Area Between Mean and Z	Z	Proportion of Area Between Mean and Z
0.40	0.1554	0.80	0.2881
0.41	0.1591	0.81	0.2910
0.42	0.1628	0.82	0.2939
0.43	0.1664	0.83	0.2967
0.44	0.1700	0.84	0.2995
0.45	0.1736	0.85	0.3023
0.46	0.1772	0.86	0.3051
0.47	0.1808	0.87	0.3078
0.48	0.1844	0.88	0.3106
0.49	0.1879	0.89	0.3133
0.50	0.1915	0.90	0.3159
0.51	0.1950	0.91	0.3186
0.52	0.1985	0.92	0.3212
0.53	0.2019	0.93	0.3238
0.54	0.2054	0.94	0.3264
0.55	0.2088	0.95	0.3289
0.56	0.2123	0.96	0.3315
0.57	0.2157	0.97	0.3340
0.58	0.2190	0.98	0.3365
0.59	0.2224	0.99	0.3389
0.60	0.2257	1.00	0.3413
0.61	0.2291	1.01	0.3438
0.62	0.2324	1.02	0.3461
0.63	0.2357	1.03	0.3485
0.64	0.2389	1.04	0.3508
0.65	0.2422	1.05	0.3531
0.66	0.2454	1.06	0.3554
0.67	0.2486	1.07	0.3577
0.68	0.2517	1.08	0.3599
0.69	0.2549	1.09	0.3621
0.70	0.2580	1.10	0.3643
0.71	0.2611	1.11	0.3665
0.72	0.2642	1.12	0.3686
0.73	0.2673	1.13	0.3708
0.74	0.2704	1.14	0.3729
0.75	0.2734	1.15	0.3749
0.76	0.2764	1.16	0.3770
0.77	0.2794	1.17	0.3790
0.78	0.2823	1.18	0.3810
0.79	0.2852	1.19	0.3830

(A)	(B)	(A)	(B)
Z	Proportion of Area Between Mean and Z	Z	Proportion of Area Between Mean and Z
1.20	0.3849	1.60	0.4452
1.21	0.3869	1.61	0.4463
1.22	0.3888	1.62	0.4474
1.23	0.3907	1.63	0.4484
1.24	0.3925	1.64	0.4495
1.25	0.3944	1.65	0.4505
1.26	0.3962	1.66	0.4515
1.27	0.3980	1.67	0.4525
1.28	0.3997	1.68	0.4535
1.29	0.4015	1.69	0.4545
1.30	0.4032	1.70	0.4554
1.31	0.4049	1.71	0.4564
1.32	0.4066	1.72	0.4573
1.33	0.4082	1.73	0.4582
1.34	0.4099	1.74	0.4591
1.35	0.4115	1.75	0.4599
1.36	0.4131	1.76	0.4608
1.37	0.4147	1.77	0.4616
1.38	0.4162	1.78	0.4625
1.39	0.4177	1.79	0.4633
1.40	0.4192	1.80	0.4641
1.41	0.4207	1.81	0.4649
1.42	0.4222	1.82	0.4656
1.43	0.4236	1.83	0.4664
1.44	0.4251	1.84	0.4671
1.45	0.4265	1.85	0.4678
1.46	0.4279	1.86	0.4686
1.47	0.4292	1.87	0.4693
1.48	0.4306	1.88	0.4699
1.49	0.4319	1.89	0.4706
1.50	0.4332	1.90	0.4713
1.51	0.4345	1.91	0.4719
1.52	0.4357	1.92	0.4726
1.53	0.4370	1.93	0.4732
1.54	0.4382	1.94	0.4738
1.55	0.4394	1.95	0.4744
1.56	0.4406	1.96	0.4750
1.57	0.4418	1.97	0.4756
1.58	0.4429	1.98	0.4761
1.59	0.4441	1.99	0.4767

(A) Z	(B) Proportion of Area Between Mean and Z	(A) Z	(B) Proportion of Area Between Mean and Z
2.00	0.4772	2.40	0.4918
2.01	0.4778	2.41	0.4920
2.02	0.4783	2.42	0.4922
2.03	0.4788	2.43	0.4925
2.04	0.4793	2.44	0.4927
2.05	0.4798	2.45	0.4929
2.06	0.4803	2.46	0.4931
2.07	0.4808	2.47	0.4932
2.08	0.4812	2.48	0.4934
2.09	0.4817	2.49	0.4936
2.10	0.4821	2.50	0.4938
2.11	0.4826	2.51	0.4940
2.12	0.4830	2.52	0.4941
2.13	0.4834	2.53	0.4943
2.14	0.4838	2.54	0.4945
2.15	0.4842	2.55	0.4946
2.16	0.4846	2.56	0.4948
2.17	0.4850	2.57	0.4949
2.18	0.4854	2.58	0.4951
2.19	0.4857	2.59	0.4952
2.20	0.4861	2.60	0.4953
2.21	0.4864	2.61	0.4955
2.22	0.4868	2.62	0.4956
2.23	0.4871	2.63	0.4957
2.24	0.4875	2.64	0.4959
2.25	0.4878	2.65	0.4960
2.26	0.4881	2.66	0.4961
2.27	0.4884	2.67	0.4962
2.28	0.4887	2.68	0.4963
2.29	0.4890	2.69	0.4964
2.30	0.4893	2.70	0.4965
2.31	0.4896	2.71	0.4966
2.32	0.4898	2.72	0.4967
2.33	0.4901	2.73	0.4968
2.34	0.4904	2.74	0.4969
2.35	0.4906	2.75	0.4970
2.36	0.4909	2.76	0.4971
2.37	0.4911	2.77	0.4972
2.38	0.4913	2.78	0.4973
2.39	0.4916	2.79	0.4974

(A) Z	(B) Proportion of Area Between Mean and Z	(A) Z	(B) Proportion of Area Between Mean and Z
2.80	0.4974	3.10	0.4990
2.81	0.4975	3.11	0.4991
2.82	0.4976	3.12	0.4991
2.83	0.4977	3.13	0.4991
2.84	0.4977	3.14	0.4992
2.85	0.4978	3.15	0.4992
2.86	0.4979	3.16	0.4992
2.87	0.4979	3.17	0.4992
2.88	0.4980	3.18	0.4993
2.89	0.4981	3.19	9.4993
2.90	0.4981	3.20	0.4993
2.91	0.4982	3.21	0.4993
2.92	0.4982	3.22	0.4994
2.93	0.4983	3.23	0.4994
2.94	0.4984	3.24	0.4994
2.95	0.4984	3.25	0.4994
2.96	0.4985	3.30	0.4995
2.97	0.4985	3.35	0.4996
2.98	0.4986	3.40	0.4997
2.99	0.4986	3.45	0.4997
3.00	0.4987	3.50	0.4998
3.01	0.4987	3.60	0.4998
3.02	0.4987	3.70	0.4999
3.03	0.4988	3.80	0.4999
3.04	0.4988	4.00	0.49997
3.05	0.4989	4.01–∞	0.5000
3.06	0.4989		
3.07	0.4989		
3.08	0.4990		
3.09	0.4990		

RESOURCE B

GLOSSARY

Ambiguous words and phrases: Confused wording that a respondent can reasonably interpret in more than one way.

Analysis of variance: A test of statistical significance for use when the dependent variable is interval and there are more than two categories of the independent variable.

ANOVA table: The summary presentation of the analysis of variance test.

Arithmetic mean: See *Mean.*

Chi-square test: A statistical significance test used for variables that have been organized into categories and presented in a contingency table.

Closed-ended questions: Questions that provide a fixed list of alternative responses and ask the respondent to select one or more of the alternatives as indicative of the best possible answer.

Cluster (multistage) sampling: The process of randomly selecting a sample in a hierarchical series of stages represented by increasingly narrow groups from the working population.

Coefficient of determination (r^2): The proportion of the variance in the dependent variable that is explained by a simple linear regression line.

Computer software packages: Prepackaged statistical programs that facilitate analysis of survey data.

Confidence interval: A probabilistic estimate of the true population mean or proportion based on sample data.

Contingency table: A tabular display presenting the relationship between two variables.

Control variable: The variable that is held constant in a three-way cross-tabulation in order to display three variables by using contingency tables.

Convenience sampling: Type of nonprobability sample in which interviewees are selected according to their presumed resemblance to the working population and their ready availability.

Correlation ratio (E^2): The proportion of the total variance that is explained by the independent variable in the analysis of variance test.

Cramér's V: A measure of association used for categorical data that is calculated directly from the chi-square statistic.

Data entry: The process of entering the raw data from completed questionnaires into a computer.

Decile deviation: The arithmetic difference between the 10th and 90th percentiles.

Degrees of freedom: Number of values in a data set that are free to vary. Once these values are known, all remaining values are also known.

Dependent variable: The variable that is being explained or is dependent on another variable.

Difference of proportions test: A test of statistical significance, with the dependent variable measured in percentages and with only two categories of the independent variable.

Direct measurement: Information-gathering technique that involves the direct counting, measuring, or testing of data.

Double-barreled questions: Questions that inappropriately elicit responses for two or more issues at the same time.

Draft questionnaire: A draft of the survey instrument that is prepared at the conclusion of the preliminary information-gathering process and prior to implementation of the pretest.

Dummy variable: The use of 0 and 1 to convert a noninterval variable into one that can be used and interpreted by regression analysis.

Emotional words and phrases: Words and phrases that elicit emotional responses rather than reasoned and objective answers.

Eta: The measure of association for the analysis of variance test.

F ratio: From analysis of variance, the ratio of the variance explained by the independent variable to the variance that remains unexplained.

Filter or screening questions: Questions that require some respondents to be screened out of certain subsequent questions or disqualified from participating in the survey at all.

Finite population correction: Adjustment of the standard error in order to account for small population sizes.

Fixed-alternative response categories: A list of response choices associated with closed-ended questions.

Focus group: A semistructured discussion among individuals who are deemed to have some knowledge of, or interest in, the issues associated with the research study.

Frequency distribution: A summary presentation of the frequency of response for each category of the variable.

Frequency polygon (line graph): Graphic tool for interval data in ordinal categories. A line connects points representing the midpoint of the class interval (horizontal axis) and the frequency of response (vertical axis).

Gamma: A measure of association for two variables, which are either ordinal or interval.

General population: The theoretical population to which the researcher wishes to generalize the study findings.

Harmonic mean: A measure of central tendency that represents the kth root of the product of the values for each category of a variable, where $k =$ the number of categories.

Inappropriate emphasis: The use of boldface, italicized, capitalized, or underlined words or phrases within the context of a question that may serve to bias the respondent.

Independent samples t test: A test of statistical significance, with the dependent variable on the interval scale and with two categories of the independent variable representing independent samples.

Independent variable: A variable that explains changes in another variable.

In-person interviews: Interviews in which information is solicited directly from respondents in a face-to-face situation.

Intercept survey: An in-person survey interview conducted in the course of the potential respondent's normal behavior pattern and without prearrangement of an interview time with that respondent.

Interquartile range: The arithmetic difference between the 25th and 75th percentiles.

Interval scale data: Data involving a level of measurement that establishes an exact value for each category of the variable in terms of specific units of measurement.

Interviewer instructions: Explicit instructions to the survey administrator concerning how to properly administer and complete the questionnaire.

Lambda: A measure of association for two variables, at least one of which is nominal in scale.

Level of confidence: Degree of confidence associated with the accuracy of the measurements derived from sample data.

Level of wording: A guideline for the development of questions that instructs the researcher to be cognizant of the population to be surveyed when choosing the words, colloquialisms, and jargon to be used in the questions.

Likert scale: A scaled response continuum measured from extreme positive to extreme negative (or vice versa) in five, seven, or nine categories.

Mail-out survey: Printed questionnaires disseminated through the mail to a predesignated sample of respondents. The respondents are asked to complete the questionnaire on their own and return it by mail to the researcher.

Manipulative information: Explanatory information in a questionnaire that is intended to provide necessary background and perspective but serves instead to bias the respondent.

Margin of error: The level of sampling accuracy obtained.

Mean (arithmetic mean): The mathematical center of the data, taking into account not only the location of the data (above or below the center) but also the relative distance of the data from the center.

Measure of association: A measure of the strength and direction of the relationship between two variables.

Measure of central tendency: Statistic that provides a summarizing number to characterize what is "typical" or "average" for particular data. Mean, mode, and median are the three measures of central tendency.

Median: The value of the variable that represents the midpoint of the data. One-half of the data will have values on one side of the median, and one-half will have values on the other side.

Mode: The category or value of the data that is characterized as possessing the greatest frequency of response.

Multiple coefficient of determination (R^2): The proportion of the variance in the dependent variable that is explained by a regression line with two or more independent variables.

Multiple regression: The statistical technique that identifies the line or curve of best fit for an interval dependent variable and two or more interval independent variables.

Nominal scale data: Data that involve a level of measurement that simply identifies or labels the observations into categories.

Nonprobability sampling: A method of sample selection in which the probability of any particular respondent's selection for inclusion in the sample is not known.

Normal distribution: Data distributed in the form of the symmetrical, bell-shaped curve, where the mode, median, and arithmetic mean have the same value.

Null Hypothesis: The hypothesis posed in tests of statistical significance and hypothesis testing indicating that whatever relationship appears to exist in the data is merely sampling error—not indicative of a real relationship.

Observation: An information-gathering technique that involves the direct study of behavior, as it occurs, by watching the subjects of the study without intruding upon them.

One-tail test: See Single- and dual-direction research hypothesis testing.

Open-ended questions: Questions that have no preexisting response categories and thereby permit the respondent to answer in his or her own words.

Ordinal scale data: Data involving a level of measurement that seeks to rank categories in terms of the extent to which these categories represent the variable.

Paired samples t test: A test of statistical significance, with the dependent variable on the interval scale and two categories of the independent variable representing samples drawn from the same population at different points in time.

Pearson's r correlation: The measure of association for two interval scale variables.

Percentile: Any percentile (k) is that value at which k percent of values in the data set are less than the kth percentile value and $(1 - k)$ percent of values are greater.

Phi: A measure of association that is a variation on Cramér's V and used only for contingency tables where at least one of the variables contains only two categories.

Post hoc tests: Multiple comparison of means tests by which the specific differences among categories of the independent variable can be identified in association with the analysis of variance test.

Postcoding: The process of coding responses to open-ended questions or other questions that are not coded as part of the precoding process.

Precoding: The placement of numeric codes for each category of response at the time that the questionnaire is prepared in final form for administration.

Pretest: A small-scale implementation of the draft questionnaire, used to assess such critical factors as questionnaire clarity, comprehensiveness, and acceptability.

Probability sample: Sample with the following two characteristics: (1) probabilities of selection are equal for all members of the working population at all stages of the selection process, and (2) sampling is conducted with elements of the sample selected independently of one another. Often referred to as *random samples.*

Proportionate reduction in error: The extent to which the independent variable serves to reduce the error in predicting the dependent variable.

Purposive sampling: Type of nonprobability sampling in which the researcher uses judgment in selecting respondents who are considered to be knowledgeable in subject areas related to the research.

q distribution: The studentized range, *q*, is a statistic used in multiple comparison methods. It is defined as the range of means divided by the estimated standard error of the mean for a set of samples being compared.

Questionnaire editing: The examination of finished questionnaires for accuracy and completeness. Often referred to as the *cleanup process.* Includes postcoding.

Quota sampling: Type of nonprobability sample in which the researcher deliberately selects a sample to reflect the overall population with regard to one or more specific variables that are considered to be important to the study.

Random-digit dialing: Use of randomly drawn telephone numbers for the purpose of contacting potential respondents.

Range: The arithmetic difference between the highest and lowest values in the data.

Regression analysis: See Simple linear regression and multiple regression.

Research hypothesis: The more common phrasing of an hypothesis that a relationship between variables—or a true finding—exists which the researcher can apply to the general population (generalize).

Respondent: The person who replies to the questions in the survey instrument.

Response rate: Percentage of the potential respondents who were initially contacted and completed the questionnaire.

Sample survey research: Survey research conducted by interviewing a small portion of a large population through the application of a set of systematic, scientific, and orderly procedures for the purpose of making accurate generalizations about the large population.

Sampling error: The likelihood that any scientifically drawn sample will contain certain unavoidable differences from the true population of which it is a part.

Sampling frame: The process of obtaining a representative sample population from the general population.

Scaled responses: Alternative responses that are presented to the respondent on a continuum.

Scatter plot: Plotting corresponding values of the independent variable (x) and the dependent variable (y) when both are interval.

Secondary research: A means of data collection that consists of compiling and analyzing data that already have been collected and that exist in usable form.

Simple linear regression: The statistical technique that identifies the line of best fit in a scatter plot of two interval variables.

Simple random sampling: The random selection of members of the working population for inclusion in the eventual sample.

Single- and dual-direction research hypothesis testing: The use of particular Z or t scores for determining statistical significance is dictated by how the research question or hypothesis is posed. Questions of difference in which the direction of difference is specified are *single-direction questions;* questions in which the direction of difference is not specified are *dual-direction questions.*

Single-sample testing (hypothesis testing): Significance tests (both interval and percentage data) comparing the results of a sample to an established standard.

Skewed distribution: A nonnormal frequency distribution with some extreme values, either high or low, that cause the three measures of central tendency to deviate from one another.

Snowball sampling: A type of nonprobability sample in which the researcher identifies a few respondents and asks them to identify others who might qualify as respondents.

Spuriousness: An apparent relationship between two variables that is found, on further analysis, to be the result of the interaction of a third variable.

Standard deviation: A measure of data dispersion that depicts how close the data are to the mean of the distribution.

Standard error: The standard deviation of a distribution of sample means, as opposed to the distribution of raw data of a single sample.

Standardized Z score: See Z score.

Stratified random sampling: Separation of the working population into mutually exclusive groups (strata); random samples are then taken from each stratum.

Student's t distribution: An adjustment to the normal distribution to account for small sample sizes.

Sum of squares: The process of summing squared differences in order to eliminate the offsetting effects of positive and negative differences.

Survey administration: The implementation of the precoded final questionnaire in the survey research process.

Survey research: The solicitation of verbal information from respondents through the use of various interviewing techniques.

Systematic random sampling: Adaptation of the random sampling process that consists of selecting sample members from a list at fixed intervals from a randomly chosen starting point on that list.

t-test: See Student's t distribution.

Telephone survey: Information collection through the use of telephone interviews that involve a trained interviewer and selected respondents.

Tests of statistical significance: A series of statistical tests that permit the researcher to identify whether genuine differences exist among variables.

Three-way cross-tabulation: The method by which the relationship among three variables is presented in a series of contingency tables.

Two-tail test: See Single- and dual-direction research hypothesis testing.

Type I error: The error associated with making a decision based on the data from the sample.

Type II error: The error associated with being overly conservative and not acting on the data from the sample.

Tukey's HSD (Honestly Significant Difference): A post hoc test for the analysis of variance.

Unit of analysis: The element (person, household, or organization) of the population that represents the focus of the research study.

Variable fields: Computer spaces allocated and corresponding to a variable in the questionnaire.

Venting questions: Questions in which the respondent is asked to add any information, comments, or opinions that pertain to the subject matter of the questionnaire but have not necessarily been addressed throughout the main body of the questionnaire.

Web-based survey: A survey administered through the Internet.

Working population: An operational definition of the general population that is representative of that population and from which the researcher is reasonably able to identify as complete a list as possible of its members.

Z score: The conversion of calculated standard deviations into standard units of distance from the mean in normal distributions.

RESOURCE C

ANSWERS TO SELECTED EXERCISES

Chapter Two

6. a. Postcodes are as follows:
   Truck driver (5)
   Veterinarian (1)
   Bookkeeper (2)
   Cashier (retail) (3) or (4)
   Attorney (1)
   City manager (1)
   Roofer (6)
   Heavy equipment operator (5), (6), or (99)
   Gas station attendant (4)
   Dancer (1) or (99)
   b. Postcode with a new category, for example, (8) Retired

Chapter Three

2. a. nominal
   b. nominal
   c. ordinal (grouped categories of interval data)
   d. ordinal
   e. ordinal
   f. nominal

g.   ordinal

h.   interval

i.   interval

j.   ordinal

3.   a.   level of wording (TANF), emotional/manipulative words (*allowed*)

b.   emotional words (*uplifting*); level of wording; ambiguous (*cultural arts*)

c.   ambiguous words/phrases (*institution, relationship*); confusing question

d.   inappropriate emphasis (*most efficient*); manipulative information (first sentence); unbalanced scale; ambiguous words/phrases (*overall efficiency*)

e.   double-barreled question (police and fire), manipulative/unbalanced question (*satisfied*)

f.   level of wording (*euthanasia*); double-barreled question (*hopelessly* or *provides consent*); ambiguous words (*practiced*); emotional words (*hopelessly*); confusing question (*below*)

g.   overlapping categories ($100,000); categories not comprehensive (limited to $200,000); ambiguous wording (*your income*); inconsistent ranges of categories

h.   categories not comprehensive; confusing question (no indication of number of responses required); ambiguous (*see*)

i.   confusing question; manipulative information (*figure skating*), not comprehensive (need more categories); ambiguous (*favorite games to watch or to participate?*)

j.   not comprehensive; manipulative information (*eliminate Saddam Hussein*), ambiguous words (*abortion, eliminate*), emotional words (*abortion*), level of wording (*sovereignty, Crusades*)

Chapter Five

1. $\bar{x} = 51.9$
median $= 49.5$
$s = 21.74$

2. $\bar{x} = \$778.11$

3.   a. $\bar{x} = 7.36$

b.   median $= 5.73$

c.   modal class $= 4$ and under 8

d. $s = 5.74$

4.   a. $\bar{x} = \$74,250$

b.   median $= \$75,417$

c. $s = \$12,920$

d. $\$85,280 - 63,750 = \$21,530$

5.   a.   median $= \$9.95$; mean $= \$11.47$

b.   interquartile range $= \$6.70$
decile deviation $= \$14.12$

On
Off

TABLE C.1.   SAT SCORES AMONG HIGH SCHOOL SENIORS.

Score	f	%
1,400–1,600	20	3.3
1,200–1,399	80	13.3
1,000–1,199	161	26.8
800–999	209	34.9
600–799	100	16.7
400–599	30	5.0
Total	600	100.0

Note: Missing cases = 50.

It would be correct to select any one of the first four categories for rounding up to 100 percent. The authors recommend selecting the category with the largest frequency.

TABLE C.2.   CITIZEN RATING OF POLICE EFFECTIVENESS.

Rating	Value	f	%
Highly satisfactory	1	140	28.0
Satisfactory	2	190	38.0
Neutral	3	75	15.0
Unsatisfactory	4	75	15.0
Highly unsatisfactory	5	20	4.0
Total		500	100.0

6. a. See Table C.1.
   b. The distribution in Table C.1 is basically characteristic of a normal distribution.
   c. mean
7. a. median
8. a. See Table C.2.
   b. Likert scale is applicable; most appropriate measure of central tendency is the mean.
   c. mean = 2.29
9. a. 324.44
   c. 311.86
   d. 85.23
10. a. $23.00
    b. $9.37
    c. $26.37

Chapter Six

1. a. .1056
   b. .9893

     c. .1977

     d. .8463

     e. .0486

     f. .1003

2.   Best performance: Test 3

    Worst performance: Test 2

3.   a. .5000

     b. .9772

     c. .2286

4.   Sarah did relatively better on the verbal section.

5.   a. .28

     b. 18.9

6.   a. 1927 New York Yankees ($Z = 2.86$)

     b. .3446

     c. 190 home runs

Chapter Seven

1.   a. $57,620–59,580

2.   49.1% –56.9%

3. $120,000 ± $2,450 = $117,550 to $122,450

4.   95 percent confidence that mean age of college students is between 25.65 and 26.75 years of age.

5. $Z = -3.84$—Yes, eligible

6. $Z = 1.94$—too much

7. $t = 1.75$—better

8.   Does not qualify sample greater than standard

9.   a. $Z = -1.12$—not different

    b. $Z = 2.44$—better

10. $Z = -2.18$—do not apply

11. $Z = -2.57$—too easy

12. $Z = -3.39$—does not meet standard

13. $t = 1.96$—yes, consistent

14. $t = 2.06$—yes, overstated

Chapter Eight

1.   No; a minimum sample size of 385 is required.

2.   a. 592

     b. 456

     c. 119

      d. 3,756
3. a. 139
      b. 373
      c. 465
4. a. 4,145
      b. 381
      c. 324
5. b. 97

Chapter Nine

2. Using the last two digits in each random number and selecting, at random, a starting place on line 4, column 2, and by reading the numbers from top to bottom, the following twenty-five numbers will be selected: 24, 37, 47, 68, 21, 67, 43, 70, 32, 17, 25, 73, 34, 13, 64, 41, 66, 45, 12, 27, 48, 60, 38, 72, and 51.

3. Begin the sample by randomly selecting a number between 1 and 2,386. Starting with the name corresponding to that randomly selected number, select every 2,386th name on the list.

4. a. White: 370
          Asian: 89
          Hispanic: 82
          Black: 59
      b. White: 300
          Asian: 100
          Hispanic: 100
          Black: 100
      c. White: 1.23
          Asian: 0.89
          Hispanic: 0.82
          Black: 0.59

5. a. Not representative
      b. Not representative
      c. Representative

7. a. Democrat   347
          Republican   83
          Independent 70
      b. Democrat   308
          Republican   96
          Independent 96

Note: When small populations yield sample sizes that are very close to the large population sample size (97), the researcher can elect to use 100 as the sample size, just as with large populations. Hence the answer could also be as follows:

Democrat: 300
Republican: 100
Independent: 100
Total: 500

c. Using the 300/100/100 distribution from Exercise 8b, weights are as follows:

Democrat   1.16
Republican   0.83
Independent   0.70

d. ±4.4 percent (95 percent confidence)
±5.8 percent (99 percent confidence)

Chapter Ten

1. a. See Table C.3.
   b. $\chi^2 = 178.227$ (significant)

TABLE C.3. POLITICAL PREFERENCE BY ETHNIC BACKGROUND.

Political Preference	White		Black		Hispanic		Asian		Total	
	f	%	f	%	f	%	f	%	f	%
Democrat	200	33.3	120	80.0	100	66.7	50	50.0	470	47.0
Republican	350	58.4	10	6.7	40	26.7	30	30.0	430	43.0
Independent	50	8.3	20	13.3	10	6.6	20	20.0	100	10.0
Total	600	100.0	150	100.0	150	100.0	100	100.0	1,000	100.0

Ethnic Background

TABLE C.4. NAFTA OPINION BY REGION OF THE UNITED STATES.

Opinion	West		Central		East		South		Total	
	f	%	f	%	f	%	f	%	f	%
Favor	70	66.7	53	44.2	100	57.1	50	50.0	273	54.6
Opposed	35	33.3	67	55.8	75	42.9	50	50.0	227	45.4
Total	105	100.0	120	100.0	175	100.0	100	100.0	500	100.0

Region

2. See Table C.4.
3. b. $\chi^2 = 60.00$ (significant)
4. $\chi^2 = 131.76$ or 124.42 (depending on recategorization); the calculated χ^2 is significant
5. a. $G = -.315$ (moderate inverse association)
    b. $Z = -1.59$ (not significant)
6. V or $f = .387$ (moderate association)
7. Depending on the recategorization employed, $V = .406$ (relatively strong association) or phi $= .558$ (relatively strong association)
8. a. $G = .38$
        $Z = 2.65$ (significant)
    b. moderate positive
9. b. $\chi^2 = 83.857$
    c. state
    d. $V = .29$ (moderate)
10. c. $\chi^2 = 37.9$ (significant)
    d. $V = .19$ (weak)

Chapter Eleven

1. $t = 2.40$ (men more in favor)
2. $t = 2.08$ (no evidence of men being older)
3. $t = 1.98$ (no proven difference)
4. $t = 1.66$ (french fry eaters consume more calories)
5. $t = 1.84$ (planners not proven to dislike jobs more)
6. $t = -1.49$ (failure not proven)
7. $t = -2.03$ (95 percent confident that his steps helped—not 99% confident)
8. a. $F = 4.432$ (yes, difference by employee category)
    b. 28.7 percent
    c. management and labor
9. b. $F = 2.85$ (no difference established at 99 percent)
10. a. $F = 10.407$ (significant difference in ages among resorts)
    b. Hawaii and Las Vegas

Chapter Twelve

1. b. $y = -1026.903 + 483.314x$
    c. $F = 14.04$ (significant)
    d. 2,840
    e. $r^2 = .637$
2. a. $y = 30.362 - .056x$
    b. $F = 19.27$ (significant)
    c. 15
    d. $r^2 = .707$

3. a. $y = 7.7 - .97x$ —4.9%
    b. $F = 5.55$ (significant)
    c. .409
    d. $r = -.64$
4. a. $r = -.346$
    b. $t = -1.043$ (not significant)
5. a. $y = .667 + 1.167x$
    b. $F = 23.059$ (significant)
    c. $r^2 = .742$
6. a. $y = 7.25 - .404x$
    b. $F = 19.279$ (significant)
    c. $r^2 = .707$
7. a. ERA line: $y = 140.793 - 13.818x$
        Batting average line: $y = 48.81 + .534x$
    b. ERA: $F = 13.933$
        BA: $F = 5.02$
        (both significant)
    c. ERA: $r^2 = .499$
        BA: $r^2 = .264$
    d. ERA of 3.90 = 87 wins
        BA of .280 = 92 wins
8. a. $y = -44.119 + 2.418x$ —predict 77 barks
    b. $F = 15.99$ (significant)
    c. $r^2 = .667$
    d. $r = .817$
    e. Table scraps stronger predictor

BIBLIOGRAPHY

Abrahamson, M. *Social Research Methods.* Upper Saddle River, N.J.: Prentice Hall, 1983.

Babbie, E. R. *Survey Research Methods.* Belmont, Calif.: Wadsworth, 1973.

Babbie, E. R. *The Practice of Social Research.* (9th ed.) Belmont, Calif.: Wadsworth, 2001.

Backstrom, C. H., and Hursh-Cesar, G. *Survey Research.* (2nd ed.) New York: Wiley, 1981.

Bailey, K. D. *Methods of Social Research.* (2nd ed.) New York: Free Press, 1982.

Bardses, B., and Oldendick, R. *Public Opinion.* (2nd ed.) Belmont, Calif.: Wadsworth, 2003.

Berk, R. A. *Regression Analysis: A Constructive Critique.* Thousand Oaks, Calif.: Sage, 2004.

Blalock, H. M., Jr. *Social Statistics.* New York: McGraw-Hill, 1972.

Converse, J. M. *Survey Research in the United States.* Berkeley: University of California Press, 1987.

Dans, J. A. *Elementary Survey Analysis.* Upper Saddle River, N.J.: Prentice Hall, 1971.

de Vaus, D. A. *Surveys in Social Research.* London: Allen & Unwin, 1986.

Devine, R. P., and Falk, L. L. *Social Surveys: A Research Strategy for Social Scientists and Students.* Morristown, N.J.: General Learning Press, 1972.

Dillman, D. A. *Mail and Telephone Surveys.* New York: Wiley, 1978.

Draper, N. R., and Smith, H. *Applied Regression Analysis.* New York: Wiley, 1966.

Glock, C. Y. *Survey Research in the Social Sciences.* New York: Russell Sage Foundation, 1967.

Greenbaum, T. L. *The Handbook for Focus Group Research.* (2nd ed.) San Francisco: New Lexington Press, 1993.

Hoinville, G., and Jowell, R. *Survey Research Practice.* Portsmouth, N.H.: Heinemann, 1978.

Kish, L. *Survey Sampling.* New York: Wiley, 1965.

Korin, B. P. *Statistical Concepts for the Social Sciences.* Cambridge, Mass.: Winthrop, 1975.

Krueckeberg, D. A., and Silvers, A. L. *Urban Planning Analysis: Methods and Models.* New York: Wiley, 1974.

Krueger, R. A. *Focus Groups: A Practical Guide for Applied Research.* (2nd ed.) Thousand Oaks, Calif.: Sage, 1994.

Levin, J., and Fox, J. A. *Elementary Statistics in Social Research.* New York: HarperCollins, 1988.

Lieberman, G. J., and Owen, D. B. *Tables of the Hypergeometric Probability Distribution.* Stanford, Calif.: Stanford University Press, 1961.

Marsh, C. *The Survey Method.* London: Allen & Unwin, 1982.

Meier, K., and Brudney, J. *Applied Statistics for Public Administration.* (5th ed.) Belmont, Calif.: Thomson Publishing, 2002.

Miller, W. L. *The Survey Method in the Social and Political Sciences.* London: Pinter, 1983.

Moser, C. A., and Kelton, G. *Survey Methods in Social Investigation.* New York: Basic Books, 1972.

Nachmias, D., and Nachmias, C. *Research Methods in the Social Sciences.* (2nd ed.) New York: St. Martin's Press, 1981.

Neuman, W. L. *Basics of Social Research.* Boston: Pearson Education, Inc., 2004.

O'Sullivan, E., and Rassel, G. R. *Research Methods for Public Administrators.* White Plains, N.Y.: Longman, 1989.

Ott, L., Mendenhall, W., and Larson, R. F. *Statistics: Tool for the Social Sciences.* (2nd ed.) Boston: PWS-Kent, 1978.

Parten, M. *Surveys, Polls, and Samples: Practical Procedures.* New York: HarperCollins, 1950.

Poister, T. H. *Public Program Analysis.* Baltimore: University Park Press, 1978.

Romer, D., and others. *Capturing Campaign Dynamics: The National Annenberg Election Survey.* New York: Oxford University Press, 2004.

Rosenberg, M. *The Logic of Survey Analysis.* New York: Basic Books, 1968.

Rossi, P. H., Wright, J. D., and Anderson, A. B. *Handbook of Survey Research.* Orlando, Fla.: Academic Press, 1983.

Schaeffer, R. L., Mendenhall, W., and Ott, L. *Elementary Survey Sampling.* (3rd ed.) Boston: PWS-Kent, 1986.

Schlaifer, R. *Probability and Statistics for Business Decisions.* New York: McGraw-Hill, 1959.

Sjoberg, G., and Nett, R. *A Methodology for Social Research.* New York: HarperCollins, 1968.

Smith, H. W. *Strategies of Social Research.* (2nd ed.) Upper Saddle River, N.J.: Prentice Hall, 1975.

Spicer, J. *Making Sense of Multivariate Data Analysis.* Thousand Oaks, Calif.: Sage, 2005.

Stewart, D. W., and Shamdasani, P. N. *Focus Groups: Theory and Practice.* Thousand Oaks, Calif.: Sage, 1990.

Sudman, S. *Reducing the Cost of Surveys.* Hawthorne, N.Y.: Aldine de Gruyter, 1967.

Sudman, S. *Applied Sampling.* Orlando, Fla.: Academic Press, 1976.

Suits, D. B. *Statistics: An Introduction to Quantitative Economic Research.* Skokie, Ill.: Rand McNally, 1963.

Warwick, D. P., and Linninger, C. A. *The Sample Survey: Theory and Practice.* New York: McGraw-Hill, 1975.

Weisberg, H. F., and Bowen, B. D. *An Introduction to Survey Research and Data Analysis.* New York: Freeman, 1977.

Welch, S., and Comer, J. *Quantitative Methods for Public Administration.* (3rd ed.) Orlando, Fla.: Harcourt, 2001.

Wilcox, R. *Applying Contemporary Statistical Techniques.* Orlando, Fla.: Academic Press, 2003.

Witzling, L. P., and Greenstreet, R. C. *Presenting Statistics.* New York: Wiley, 1989.

Wolf, F. L. *Elements of Probability and Statistics.* New York: McGraw-Hill, 1962.

Yamane, T. *Statistics: An Introductory Analysis.* (2nd ed.) New York: HarperCollins, 1967.

Young, P. V. *Scientific Social Surveys and Research.* Upper Saddle River, N.J.: Prentice Hall, 1966.

INDEX